GIS: A Sourcebook for Schools

Edited by
David R. Green

London and New York

First published 2001 by
Taylor & Francis
11 New Fetter Lane, London EC4P 4EE

Simultaneously published in the USA and Canada
by Taylor & Francis Inc
29 West 35th Street, New York, NY 10001-2299

Taylor & Francis is an imprint of the Taylor & Francis Group

© 2001 David R. Green except chapter 4 © ESRI

Typeset in Sabon by
HWA Text and Data Management, Tunbridge Wells
Printed and bound in Great Britain by
TJ International Ltd, Padstow, Cornwall

All rights reserved. No part of this book may be reprinted or
reproduced or utilised in any form or by any electronic, mechanical, or
other means, now known or hereafter invented, including photocopying
and recording, or in any information storage or retrieval system,
without permission in writing from the publishers.

Every effort has been made to ensure that the advice and information in
this book is true and accurate at the time of going to press. However,
neither the publisher nor the authors can accept any legal responsibility
or liability for any errors or omissions that may be made.

British Library Cataloguing in Publication Data
A catalogue record for this book is available from the British Library

Library of Congress Cataloging in Publication Data
Green, David, R., 1956–
    GIS : a sourcebook for schools / David Green.
        p. cm.
    Includes bibliographical references.
        1. Geographic information systems. 2. Geography–Study and
    teaching (Secondary)–Audio-visual aids. I. Title.

G70.212 .G74 2000   00-041754
910'.285–dc21

ISBN 0–7484–0270-5

QUEEN MARY
UNIVERSITY OF LONDON
LIBRARY

Contents

Tables

Figures

Contributors

Ove Biilmann, 19 Anemonevej, DK 2970 Horsholm, Denmark. Tel: +45 42869568

Ralf Bill, Department for Land Cultivation and Environmental Protection, Institute for Geodesy and Geoinformatics, Rostock University, D-18059 Rostock, Germany. Tel: +49 381 498 2185, fax: +49 381 498 2188. E-mail: ralf.bill@agrarfak.uni-rostock.de. Internet: http://www.agr.uni-rostock.de/gg/

Michael J. Brown, Coastal Studies and Technology Center, Seaside High School, 1901 N. Holladay Drive, Seaside, OR 97138, USA. Tel: +1 503 738-5586. Fax: +1 503 738-5589. E-mail: mbrown@seaside.k12.or.us. Internet: http://www.seaside.k12.or.us.

Angela Crechiolo Best, Department of Geography, Environmental Sciences Building, Trent University, 1600 West Bank Drive, Peterborough, Ontario K9J 7B8, Canada. Tel: +1 705 748-1011 ext 1571. Fax: +1 705 748-1205. E-mail: angiebest@trentu.ca. Internet: http://www.trentu.ca/academic/geography/abbio.html.

Charlie Fitzpatrick, ESRI Schools and Libraries, 1305 Corporate Center Drive, Suite 250, St. Paul, MN 55121-1204, USA. Tel: +1 651 994 0823 ext.8349, Fax: +1 651 454 0705. E-mail: cfitzpatrick@esri.com. Internet: http://www.esri.com/k-12

Stephen Gill, Powys County Council, County Hall, Llandrindod Wells, Powys, LD1 5LG, United Kingdom. Tel: +44 1597 827582. E-mail: steveg@powys.gov.uk

Michael Gould, Lenguajes y Sistemas Informáticos, Universitat Jaume I, E-12080, Castellon, Spain. Tel: +34 964 72 83 17. Fax: +34 964 72 84 35. E-mail: mail@mgould.com.

David R. Green, Centre for Remote Sensing and Mapping Science, Department of Geography, University of Aberdeen, Elphinstone Road, Aberdeen, AB24 3UF, Scotland. Tel: +44 1224 272324, fax: +44 1224 272331. E-mail: d.r.green@abdn.ac.uk. Internet: http://www.abdn.ac.uk/geospatial/

Stewart McCall, GIS Co-ordinator, Information Services, South Ayrshire Council, County Buildings, Wellington Square, Ayr KA7 1DR, United Kingdom. Tel: +44 1292 612733. Fax: +44 1292 612402. E-mail: stewart.mccall@south-ayrshire.gov.uk

David J. Maguire, Managing Director, ESRI (US), 380 New York St, Redlands, CA 92373-8100, USA. Tel +1 909 793-2853. E-mail: dmaguire@esriuk.com

Mark Oliver, GIS Co-ordinator, Napanee District Secondary School, Napanee, Ontario, Canada. Tel: +1 613 354 3381 ext. 324, Fax: +1 613 354 1826. E-mail: markoliv@mercury.kosone.com

Jim Page Database Business Services, Ordnance Survey, Romsey Road, Southampton, SO16 4GU, United Kingdom. Tel: +44 23807 9243. E-mail: jpage@ordsvy.org.uk.

David Rhind, Vice Chancellor, City University, Northampton Square, London EC1V 0HB, United Kingdom. Tel: +44 207 477 8002. E-mail: drhind@city.ac.uk.

Peter Roberts, Powys County Council, County Hall, Llandrindod Wells, Powys, LD1 5LG. Tel: +44 1597 827582. E-mail: peterr@powys.gov.uk

Bob Sharpe, Department of Geography and Environmental Studies, Wilfred Laurier University, Ontario, Canada, N2L 3C5. Tel: +1 519 884 1970 ext. 2684, fax: +1 519 725 1342. E-mail: bsharpe@mach1.wlu.ca

Stephen Walker, Department of Geography, The Holgate School, Hillcrest Drive, Hucknall, Nottinghamshire NG15 6PX, United Kingdom. Tel: +44 115 963 2104. Fax: +44 0115 968 1993. E-mail: stephen-walker@lineone.net

Gwil Williams, Humanities Co-ordinator, Horndean Community School, Barton Cross, Horndean, Hampshire PO8 9PQ.

Abbreviations

3-D	three-dimensional
ADE	autodesk data extension
AEGIS	An Educational Geographic Information System
AGI	Association for Geographic Information
AOL	America Online
ASCII	American Standard Code for Information Interchange
ATM	asynchronous transfer mode
AUME	Advisory Unit for Microtechnology in Education
BBC	British Broadcasting Corporation
BT	British Telecom
CAD	computer-assisted drafting or computer-aided design
CEO	chief executive officer
CIA	Central Intelligence Agency
CLI	Command Line Interface
CREST	Columbia River Estuary Study Taskforce
DBMS	database management system
DSS	Decision Support System
EASICC	Education and Awareness Special Interest Conference Committee
EIS	environmental information systems
ESRC	Economic and Social Research Council
ESRI	Environmental Systems Research Institute
FHS	polytechnic schools (Germany)
FTP	file transfer protocol
GA	Geographical Association
GADS	Geographic Analysis and Display System
GAN	Geographic Alliance Network
GCSE	General Certificate of Secondary Education
GIS	Geographical Information Systems
GISAS	GIS Applications Specialist Program
GIST	GIS Tutor
GMS	grant maintained schools

GNVQ	General National Vocational Qualification
GPS	global positioning system
HTML	hypertext markup language
I&E	Information and Education
IFB	Internet for Business
INSET	in-service training
IT	information technology
ITC	International Institute for Aerospace Survey and Earth Sciences (the Netherlands)
JPEG	Joint Photographic Expert Group
KS1	Key Stage 1
lab.	laboratory
LANs	local area network
LAs	Local Authorities
LEA	Local Education Authority
LMS	local management of schools
MA	Mapping Awareness
MNR	Ministry of Natural Resources
NATO	North Atlantic Treaty Organisation
NCGE	National Council for Geographic Education
NCGIA	National Center for Geographical Information Analysis
NDSS	Napanee District Secondary School
NGS	National Geographic Society
NIESU	Northern Ireland Education Support Unit
NRSC	National Remote Sensing Centre
NT	New Technology
NWI	National Wetlands Inventory
OAC	Ontario Academic Courses
OAGEE	Ontario Association for Geographic and Environmental Education
OCR	optical character recognition
ODFW	Oregon Department of Fish and Wildlife
OS	Ordnance Survey
OSNI	Ordnance Survey of Northern Ireland
PC	personal computer
PCC	Powys County Council
PR	public relations
RAM	random access memory
RM	Research Machines
RRP	recommended retail price
RS	remote sensing
SAS	statistical analysis software
SEP	Secondary Education Project
SIT	spatial information technologies

SLA	service level agreement
SPOT	système probatoire de l'observation de la terre
SQL	structured query language
SPSS	statistical package for the social sciences
SSFC	Sir Sanford Fleming College
SSSI	sites of special scientific interest
TEC	Training and Enterprise Council
UK	United Kingdom
UNESCO	United Nations Educational, Scientific and Cultural Organisation
URL	uniform resource locator
USA	United States of America
USGS	United States Geological Survey
UTM	universal transfer mercator
VRML	virtual reality markup language
WAN	wide area network
WLU	Wilfrid Laurier University
WWW	World Wide Web

Preface

David R. Green

The initial reason for writing this short book was to provide a basic reference on Geographical Information Systems (GIS) in School Education (elementary, primary and secondary) for both teachers and pupils, thereby filling an apparent gap on the current list of GIS and geography textbooks. The text also gathers together some of the material under 'one roof', material that might not otherwise reach the intended audience.

The starting date for this project goes back many years now but, whilst quite a few exciting developments have taken place since the commissioning of this book by Taylor & Francis, the topics covered by the individual chapters here are still as relevant to GIS in school education today as they were a few years ago, especially in the UK.

The general interest on my part in seeking to extend the teaching of GIS in the education system, by introducing the topic of Geographical Information Systems (GIS) into the school curriculum, encouraged me to take things a little further. At the outset the easiest way to achieve this goal was to pull together some of the then current research and examples which had begun to appear into a number of AGI (Association for Geographic Information) GIS education sessions and subsequently into various different conference proceedings both in the UK and Europe. The main aim of these early conference sessions and papers was to try and stimulate both interest in teaching GIS in schools and much needed channels of communication and links with those parties who might be able to help, from a number of different perspectives: industry, academia, commerce, and hopefully even school teachers. Whilst this was initially successful to some degree and succeeded in raising awareness, such conference publications did not reach too many school teachers. In many ways the initiative really came too soon.

Although this publication comes quite a number of years later, the wait would, however, from my perspective, seem to have been more than worthwhile, as many developments in GIS and primary and secondary education, both in the UK and around the world, have led to:

- more publications on GIS in the school environment, particularly originating in North America

- the development of teaching resource packs
- demonstration and GIS software and CD-ROMs
- greater involvement from commercial software and data suppliers
- interest from mapping agencies e.g. the Ordnance Survey, and the Ordnance Survey of Northern Ireland
- interest from a number of Local Authorities (LAs) in the UK regarding the provision of data
- a growing number of projects and theses on GIS in schools as an outcome from GIS degree programmes at both the undergraduate and postgraduate level
- greater involvement from teachers in using the technology
- commercial sponsorship of school GIS prizes e.g. the Environmental Systems Research Insititute (ESRI) (UK)
- greater interest from GIS bodies e.g. the Association for Geographic Information (AGI) via e.g. the AGI Information and Education Committee
- growing awareness of the need to make greater use of information technology (IT) in schools

and last but not least

- the development of the World Wide Web (WWW) or Internet

All of the above developments have greatly improved the opportunities, awareness and interest in GIS at the school level, even in elementary and primary school education.

The outcome of all these developments therefore is hopefully a book which offers both teachers and pupils alike a definitive source of information about GIS in schools: what it is, where it originated, what it involves, and some useful practical examples developed by practising teachers of the applications currently being used in the school classroom. The target audience has meant that there has been a requirement to provide such information at a level which is relatively easy to grasp and easy to digest. It has also been important to convey the technical nature of GIS in such a way that explains the power of the 'data handling toolbox' available in the context of 'real-world' applications, but without swamping the reader with details and information that are more appropriate for students of both further- and higher-education GIS courses. Although this is a difficult task, bearing in mind for a moment the complexity of GIS technology, the links to geography, cartography (maps) and IT have helped to provide a starting point upon which examples can be used to illustrate the functionality in a simple, understandable, and yet effective manner. As a result the chapters, to a large extent, have been written by enterprising school teachers and school educators who have daringly taken the 'plunge' into the 'world of GIS' by recognising the potential of this tool for both teaching about the technology but also about geography and other areas of the curriculum that include the use of spatial data and information. Additionally, some of the authors are people who have had close ties with school education and have sought to communicate this potential to a new audience.

In some ways this has been made quite difficult, as GIS technology has frequently been perceived to be very new, very technical and very demanding in terms of the IT learning curve required, and I suspect therefore misunderstood by many people. In part the problem of introducing this subject into a more traditional curriculum has also been made more difficult as a result of there being few teaching support materials available, and a lack of suitable software and hardware.

As part of the 'educational continuum', I also felt that it was necessary to add several chapters that examined GIS education from the perspective of institutions of both further and higher education. This has the benefit of offering teachers and older pupils alike a better insight into the role and importance of GIS in the real world, and also to provide some useful pointers towards the pursuit of more education in this rapidly growing field.

To extend the range of material available to schools, it is my intention, (to coincide with this book), to provide a new website located at the University of Aberdeen which will focus solely on GIS education in schools. The site will have the following WWW address:

http://www.abdn.ac.uk/gis_school/

and will initially comprise:

- a **point of contact**
- a series of **downloadable documents** (in Word and Acrobat® Reader™ format) – taken from the GIS in Education articles in Mapping Awareness and Mapping Awareness and GIS in Europe
- **software demos** (downloadable GIS demonstrations)
- **links** to other GIS in Schools web pages (including, for example, the AGI, and ESRI, GRID-Arendal, IDRISI (Clark Labs))

In addition, two items of software from ESRI will be provided to accompany this text.

Personal experience has shown that GIS technology has gradually filtered down from the higher education institutions into primary and secondary education, largely through the efforts of researchers and academic staff in universities connecting with teachers in schools in the locality of a higher education institution. The origins of many GIS courses in geography departments, at least initially, has also meant that contacts with geography departments in schools have been a logical starting point for both academics and teachers alike. Indeed some of the early publications show that contact between some schools and higher education has been very positive and has led to a starting point for developing basic teaching materials to aid school teachers. A small committee, established by the Geographical Association (GA), eventually led to a number of short papers being published in *Teaching*

Geography (Freeman et al., 1993; 1994). These were a useful start for the development of further school contacts, and were soon followed in later years by GIS sessions at the annual GA Conference, one of which included a presentation from Professor David Rhind (ex-Director General of the Ordnance Survey, UK).

When I started to become interested in GIS in school education, the initiative appeared to be coming from a small number of academics who had initially recognised the importance and relevance of GIS to schools in the context of geography. With growing interest gradually being shown by a number of school teachers, awareness of GIS started to grow, the end result being the development of simpler GIS software and resource packs designed to help the school teacher start to get to grips with GIS, to see its potential, and to provide ideas and feedback.

In addition, with the relatively recent expansion of the Internet and the World Wide Web (WWW), all-pervasive technology that has migrated from the relative isolation of the academic community to the workplace and into the home, GIS too has, not surprisingly, become an Internet-based technology. Access to this combination of multimedia, database, communications, and networking technology has grown at a phenomenal rate, and with the rapid development of computer processor technology, has now provided an opportunity to develop 'on-the fly' map creation and delivery systems. As this technology has become cheaper, and more schools have been linked to the Internet, there is once again yet another opportunity for school-age children and teachers to not only gain access to this vast information resource, but also to interact directly with GIS technology. This opportunity will no doubt be further enhanced with the recent announcement by the British Prime Minister, Tony Blair, of a £700-million to £1-billion investment in computers, Internet technology and training for schools in the UK over four years (Carvel, 1998).

It is therefore perhaps now an opportune moment in time to introduce this book to schools, in the hope that even more people will soon begin to recognise the growing importance of this technology in the world outside the classroom. Also as an opportunity to make future generations more aware of the frequency with which we use geographic data and information in our daily lives within a computer environment as the basis for access to information, planning, and often as a Decision Support System (DSS).

As the posters in the recently available ESRI GIS Teachers Pack (included as part of a recent GIS Awareness week in the United States) succinctly and aptly state:

- 'People, places and patterns: geography puts the pieces together'
- 'Making connections between maps: discovering Geographic Information Systems'
- 'Explore your world with GIS'
- 'Geography matters'

References

Carvel, J. (1998). Computer Revolution will Connect Schools to a National Grid for Learning, Preventing Emergence of an 'Information Poor' Generation: Blair's £1bn IT Package for Schools. *The Guardian*, Saturday, November 7th 1998, Home News, p. 7.

Freeman, D., Green, D., Hassell, D. and Paterson, K. (1993). Getting Started with GIS. *Teaching Geography*, April 1993, pp. 57–60.

Freeman, D., Green, D. and Hassell, D. (1994). A Guide to Geographic Information Systems (GIS). *Teaching Geography*, January 1994, pp. 36–37. Information Technology Section.

Acknowledgements

I would like to acknowledge the help of Richard Steele of Taylor & Francis without whom this book would never have seen the light of day. More recently, I would like to acknowledge the help of Luke Hacker and Tony Moore, also of Taylor & Francis, who have constantly kept the pressure on. It also goes without saying that I would like to thank all those people who contributed to this volume, albeit a long time in going to print! Jenny and Alison of the Department of Geography Cartography Unit at the University of Aberdeen were responsible for drafting and re-drafting many of the figures used in the text.

Over the years a number of other people have both added to and enriched my knowledge and enthusiasm through discussions about GIS in school education, and I would like to take this opportunity to thank them all. These include: David J. Maguire, Charlie Fitzpatrick and Roy Laming (ESRI (US) and ESRI (UK) respectively); Jack Dangermond (ESRI (US)) for allowing his staff to co-operate so freely and efficiently; Ian Heywood (now The Robert Gordon University, Aberdeen); Karen Kemp (now University of Redlands, California); Seppe Cassettari (now The GeoInformation Group); David Rhind (now City University); David Unwin (now Birkbeck College); Chris Coggins (now The University of Sheffield); Steve Palladino and Mike Goodchild (now Ventura College and NCGIA respectively); Pip Forer (University of Auckland, New Zealand); Fred Toppen (University of Utrecht); Lindsay McEwen (The College of St. Paul and St. Mary, Cheltenham); Diana Freeman (Advisory Unit: Computers in Education); Rob Wheatley (Langdon Park School); Christine Warr; Roy Newell (AGI); and most recently Joseph Kerski (USGS), Chris Corbin (AGI), and Judith Mansell (Education Officer) of the RGS (Royal Geographical Society (with IBG), London). The work has also had support from the Association for Geographic Information (AGI) through: allowing me to run a number of 'GIS in Education' Sessions at past AGI conferences; giving me permission to reproduce the AGI Glossary of GIS Terms (courtesy of Gayle Gander and Shaun Leslie); being involved in the AGI Information and Education Committee (I&E); and latterly for letting me set up a GIS in

School Education AGI Special Interest Group (www.gis-education.com) with the help of Stephen D. King, Gayle Gander and Kritee Apajee.

I would also like to thank a very dear friend and colleague, Stephen D. King (University of Aberdeen) who, besides reading and commenting on my chapters, has provided considerable encouragement and advice in the past year which has helped to get this book into print. Last, but not least, my mother Mrs E.G. Green.

David R. Green

Chapter I

GIS in school education

An introduction

David R. Green

A number of years ago, eight to be exact, I started a monthly column in what was then *Mapping Awareness* (MA) and subsequently became *Mapping Awareness and GIS in Europe* on 'GIS in School Education'. This was carried on by colleague Mark Chaloner. At that time in the UK, relatively little consideration had been given to the teaching of Geographical Information Systems (GIS) in the school curriculum, although a number of papers, presented at special sessions of the annual conference of the Association for Geographic Information (AGI), had discussed various different aspects of introducing GIS into school education and the school curriculum, as well as higher education: for example, the AGI Conference and Exhibition of 1991 and 1992 (Cassettari, 1991; Clark, 1991; Freeman, 1991; Green and McEwen, 1992; Heywood and Petch, 1991; Kemp, 1991; Langford and Strachan, 1991). The AGI Yearbook (1991) saw the inclusion of some chapters on GIS in Education (Green and McEwen, 1991; Unwin and Dale, 1991). Freeman (1993) produced an AGI monograph entitled 'GIS in Schools'. The 1992/93 and 1994 AGI Sourcebooks also included a number of papers on GIS in higher and secondary education (Forer, 1994; Gittings et al., 1993; Green, 1993, 1994; Green and McEwen, 1994; Maguire, 1993; Masser and Toppen, 1994; Palladino, 1993a, 1993b, 1994a, 1994b; Petch et al., 1994; Raper, 1993; Smith, 1994; Vicars, 1993; Wheatley, 1994; Wood and Cassettari, 1993).

Advisory Unit for Microtechnology in Education

It was about this time that the GIS software, AEGIS, designed specifically for schools was first marketed, by the Advisory Unit for Microtechnology in Education (AUME, 1990) (now Advisory Unit: Computers in Education). Although not a fully functional GIS package as such, AEGIS offered an essential introduction to GIS for both pupils and teachers alike, a relatively rare software resource at the time. Designed by school educators to run on relatively low-level computer systems, the sort that would typically be found in schools of the time, AEGIS provided both teachers and pupils with a new type of geographical experience. Developed for schools, this system combined a mapping and database package for MS-Windows on the RM-(Research Machines) Nimbus platform as well as on RISC-OS for the Acorn Archimedes/A3000/A5000 series, two computer platforms widely popular in UK schools at the time. Although geographical in its origin, the software was also designed to be used across the curriculum where maps and diagrams were employed. The software came with sample data files and maps, technical documentation, and materials designed to support the UK National Curriculum in Key Stages 2, 3 and 4.

Mapping Awareness and GIS in Europe

An approach aimed at circulating Mapping Awareness and GIS in Europe at a reduced rate into schools was initiated in 1991 so as to ensure that (a) a valuable link was established between education and the commercial world responsible for providing the necessary computer software and data, and (b) teachers could get access to the monthly 'GIS in Education' column and look more widely at the rest of the Mapping Awareness magazine to provide background information and insight into the developing GIS technology. This was primarily a 'raising awareness' exercise which sought to attract attention and to draw the teaching community into the world of GIS.

Whilst both of the above developments served their purpose, in many respects the promotion of GIS in school education was perhaps a little too soon – 'before its time' you might say – and a good number of years were to pass before the profile of GIS in school education was seemingly raised to a reasonable level.

NCGIA

In the United States (US) things were generally a little more successful and progress was more rapid with respect to GIS in schools. In part this was a result of the general GIS and educational activities of the National Center for Geographical Information Analysis (NCGIA) (see e.g. Palladino and Kemp, 1991). The NCGIA initiated a school-oriented GIS programme as part of its remit, some details of which are provided in Palladino (1993a), and Palladino and Goodchild (1993). The early stages of this programme (Palladino, 1993a) were reported in the Yearbook of the AGI Geographic Information 1992–1993 (Cadoux-Hudson and Heywood, 1993), and subsequent developments (Palladino, 1994a) in a later The Association for Geographic Information 1994 Sourcebook of GIS (Green et al., 1994). Further reports from the NCGIA include Palladino (1994b) and Palladino and Van Zuyle (1996).

Reference to the NCGIA Home Page (http://www.ncgis.ucsb.edu/) also reveals that the schools programme has been an on-going process and by 1993 had culminated in a special meeting for school teachers.

The NCGIA Secondary Education Project (SEP) (NCGIA, 1993a, 1993b) aimed to:

- facilitate dialogue between interested parties
- collect, develop and review GIS instructional materials
- pool information on current and pending GIS activities in schools
- communicate information
- provide specific activities that increase teacher and student awareness of GIS
- encourage development of appropriate GIS instructional materials for schools

One of their first activities was to develop a one-week GIS workshop for high-school teachers. Subsequently this developed into a series of 'GIS in the Schools' workshops. These provided:

- a short course in GIS
- observation of working GIS applications
- hands-on work with GIS software
- discussion of the role of GIS in the schools
- a review of available instructional materials

To broaden awareness, it was not only GIS teachers who were invited to participate. In addition, representatives from the National Geographic Society (NGS) sponsored Geographic Alliance Network (GAN) were invited to attend and to give presentations. The SEP Phase II led to the development of instructional materials to meet demand by school teachers. The outcome of the work was the 'GIS in Schools' Workshop Resource Packet. The aim of the 'packet' was to aid institutions with GIS expertise and technology in their efforts to create outreach activities for their local schools, with additional possible use in secondary schools. The Resource Packet included:

- an outline of the workshop components
- an evaluation of the success of the components
- suggested format for the future workshops
- a section reviewing the prospects for GIS in secondary schools
- a set of teacher project summaries (mainly manual activities that could be dated to GIS software)
- GIS short-course outline notes
- GIS for schools resource list (software and curriculum details)
- GIS glossary of terms

A package known as GADS – Geographic Analysis and Display System developed by PhD Associates has also been used in teaching in several schools in Ontario, Canada (NCGIA, 1993a). The software comes with a world statistics dataset and two lesson plans:

- Lesson 1: analysis of land use and settlement patterns using a geographic base map with elevation information
- Lesson 2: analysis of the ozone hole using ozone measurements in the upper atmosphere to create maps of ozone concentration in the polar regions.

Various other attempts to develop and promote GIS in school education have been reported by the NCGIA (NCGIA, 1993b). These include papers given at the National Council for Geographic Education (NCGE) Annual Meeting in Halifax, Nova Scotia (August 1993). The draft version of the National Geography Standards outlines the content and performance objectives for future K-12 geography courses intended to satisfy the National Educational Goals in Geography for the year 2000. In this document GIS is to be both a topic of study (database with mapping capabilities) and a support for the exploration of various geographical concepts.

A GeoSIM module developed at Virginia Tech. for undergraduate GIS courses has also been considered as a possible resource for school education. Similarly a package called GeoMedia (not to be confused with the Intergraph GeoMedia product) from the United States Geological Survey (USGS) has been considered as a possible precursor to GIS use in Higher Grades.

Geographical Association

Nearly all of the published papers available early on were authored by individuals from institutions of higher or further education, rather than by school teachers. To try and reach the teaching community, several attempts were made to raise teacher awareness in UK schools. The first approach involved the writing of a series of short introductory papers for inclusion in the Geographical Association (GA) journal, *Teaching Geography* (Freeman et al., 1993, 1994). The aim of this series of papers was to describe what GIS are, how they are used, and where they fitted into the school geography curriculum, bearing in mind the relatively recent inclusion of the term GIS in the school curriculum at the time. Besides short paragraphs of text, the papers, spanning a single page, included a selection of diagrams designed to illustrate some key components of GIS.

In the UK, Dr Chris Coggins (then at Luton College of Higher Education and now at the University of Sheffield), Chair of the Geographical Association Working Group on GIS in School Education, developed a school curriculum designed to complement the one produced earlier for higher education by Professor David Unwin (then at the University of Leicester, and now at Birkbeck College, London) (Unwin, 1990).

Association for Geographic Information

Interest in the teaching of GIS in school education in the UK has also progressed through the work of the Information and Education (I&E) Committee of the Association for Geographic Information. Membership of this committee included a number of individuals who were either involved directly with school education, e.g. Diana Freeman, An Educational Geographic Information System (AEGIS), or who had a strong interest in school education (e.g. David R. Green, David J. Maguire, Roy Newell). A publication by Freeman (1993) entitled 'GIS in Schools' was published by the AGI and is available for free. Over the years members of the AGI I&E Committee have continued to maintain strong links with the Geographical Association, and in particular have ensured that the AGI has a presence at its annual conference. Additionally, individuals such as Professor David Rhind (ex-Director General and Chief Executive of the Ordnance Survey (OS)) have continued to play a part in the promotion of links between the OS, the GA and the AGI (Rhind, 1993a, 1993b). It is likely that the work of the AGI, in embracing a 'broad church', will continue to reach out to the educational community at large. In the context of schools, as Newell (1997) notes, however, 'there is still much to be done as general awareness of GIS ... on the whole is still generally lacking' (p. 12). More recent details concerning the activities of the AGI in school education can be found in AGI (1997).

ESRI (UK) and ESRI (US)

The Environmental Systems Research Institute (ESRI) established by Jack Dangermond in 1969 has, largely through the efforts of individuals such as David J. Maguire (ESRI [UK]) and Charlie Fitzpatrick (ESRI [US]), built up very strong links with schools through the creation of an ESRI Schools Prize in the UK and the US. This initiative has been very successful, particularly in the US, and has further helped to extend a 'friendly arm' to UK schools (both teachers and pupils) with the aim of introducing them to GIS initially through the ArcView software, and latterly ArcExplorer, together with other ESRI and related products.

ESRI in the US has continued to be very active in promoting GIS in schools, and to this end in 1993 came up with *ArcSchool Reader*, a newsletter for people interested in promoting GIS education through elementary and secondary schools and the library world (Fitzpatrick, 1993). Besides news stories, this newsletter included instructions for using ArcView with ArcUSA, ArcWorld, and ArcScene. A specifc product highlighted in the first issue was the ArcView Coloring Book designed to provide guidance for those creating displays with ArcView and ArcData. Summaries of how ArcView might be used in various disciplines and grade levels in kindergarten through to 12th grade were also included. As much as anything else this newsletter offers a resource and includes contacts and addresses for any interested parties.

Links between ESRI and schools have continued to develop, and most recently they have released a Schools Package, together with a couple of new CD-ROMs. This latest package provides schools with an extremely useful resource for teaching which takes advantage of the latest ESRI technology. This package, available from Charlie Fitzpatrick of ESRI's US office, provides the following:

- A CD-ROM – GIS for Schools and Libraries – Version 4 (Educational CD about geographic information systems for Macintosh and Windows)
- A CD-ROM – Geography Awareness Week 1998

Both CDs provide software and geographical datasets which when taken together offer both an exciting and informative look at GIS, making use of GIS software that has been effectively 'geared down' to suit the specific end-user market – in this case ArcView in the form of ArcExplorer (see figure 1.1).

The approach of offering 'cut-down' software with an appropriately simplified user-interface is, in fact, exactly what both a colleague and myself had in mind a number of years ago in the articles we wrote for *Mapping Awareness* (Green, 1991a, 1991b, 1992a, 1992b; Chaloner, 1992a, 1992b, 1992c, 1992d). The package also includes a teacher resource pack, and a number of posters. There is no doubt in my mind that such products are an important

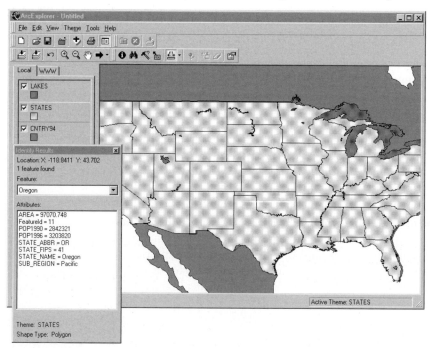

Figure 1.1 ArcExplorer (Graphic image supplied courtesy of Environmental Systems Research, Inc. and is used herein with permission)

way of getting the message about GIS across to both teachers and the up-and-coming generations, and offering the package in this form provides a superb and very visually vibrant and practical way of making the geographic technology all that more interesting. Such packages are also assisted by articles such as those provided in the ESRI newspaper *ARC*.

More recently with the development of the Internet, ESRI has taken the lead in many respects by providing access to downloadable GIS software and datasets on its website (http://www.esri.com/) on a set of special 'schools' pages (see also ESRI, 1993, 1997, 1998a, 1998b). With growing access to the Web in schools, this is a unique and very practical initiative which should help teachers and pupils to make a start on the GIS front.

TYDAC SPANS®

In Canada, collaboration between the GIS software developer TYDAC SPANS® and the Napanee District Secondary School led to a GIS education project (SPANS®, 1995). Started in 1995, Ontario's Ministry of Education and Training and the Lennox and Addington County Board of Education began a project which sought to combine GIS skills with a teacher-training model and curriculum toolkit. TYDAC provided the project with SPANS® Map, SPANS® GIS and SPANS® Explorer as the software base. The project co-ordinator Mark Oliver (see chapter 6), worked with TYDAC to develop new geography classes in which students learn a variety of GIS skills and see some example applications such as mapping projects for a community police association, Regional Conservation Authorities, and private businesses. One very positive outcome of the project was the opportunity for student work placement, e.g. the Ministry of Natural Resources, as well as the greatly increased self-esteem and proficiency in GIS skills acquired by students. Most notable is the observation concerning recognition of the importance between geography and work skills required by employment, and the feasibility and practicality of 'teaching' GIS skills within the secondary school curricula. Other work in Ontario has also been carried out by Sharpe and Crechiolo (1999). See also Autodesk (1997a; 1997b).

IDRISI (Clark Labs)

Besides both ESRI and TYDAC's involvement in school education, the IDRISI software package from Clark Labs in the USA has also proved very successful as a GIS toolbox in schools. Originally a DOS-based system with a Command Line Interface (CLI), but now running under MS-Windows (3.1, 95, 98, 2000, NT), IDRISI has also gradually found its way into schools through special licensing agreements (Summer 1995 issue of newsletter) which facilitated some 250 copies of IDRISI being shipped to both primary and secondary educational institutions around the world (IDRISI, 1996). For a low cost of approximately $100.00 schools could purchase the software for five machines. To aid teachers who were involved in introducing GIS (and image processing – IDRISI is an integrated GIS and image-processing software package) into the classroom, the IDRISI Project hosted a three-day teacher-training workshop (August 1996) for some ten teachers from the US and Canada. A similar project was held in the summer of 1997. In the spring of 1997, a teacher's pack of exercises aimed at high-school level education was released. See also IDRISI (1996) and IDRISI (1998), and http://www.clarklabs.org.

The Ordnance Survey (OS)

More recently, further interest in the development of GIS in the school curriculum has been revealed through the work of the Ordnance Survey, the UK national mapping agency, and the Association for Geographic Information. In the UK, the Ordnance Survey announced a new schools-related initiative described in Smith (1994). Collaboration between Local Authorities, schools and the Ordnance Survey is described in Gill and Roberts (1998). Details on OS and GIS can be found in Ordnance Survey (1999). Rhind (1993a) has also written about the relationships between maps, information and geography, and Ordnance Survey and schools (Rhind, 1993b).

The Ordnance Survey of Northern Ireland (OSNI)

The Ordnance Survey of Northern Ireland has also developed a CD-ROM 'Geographic Information Systems for Schools' (produced by the Ordnance Survey of Northern Ireland in conjunction with the Northern Ireland Education Support Unit and the Education and Library Boards). A paper by Brian Galloway (1997) details the work of the Ordnance Survey for Northern Ireland (figure 1.2). The CD-ROM offers a structured programme of education for both the school teacher and the school pupil. In September 1996, GIS became part of the Key Stage 4 GCSE geography curriculum requirement in Northern Ireland (Galloway, 1996). In a joint venture, the Ordnance Survey of Northern Ireland (OSNI) has worked in conjunction with the Northern Ireland Education Support Unit (NIESU) and a number of others, e.g. ESRI, to develop the materials needed to address the new GIS teaching and learning needs. The end product, an interactive multimedia CD-ROM, offers a mechanism by which it is possible to raise awareness through the introduction of maps and GIS, ranging from an introduction to the subject matter through to coverage of GIS functionality and even applications that allow pupils to input their own data. It is also possible for both teachers and pupils to make notes using the notebook button. Additional help files are also provided, as well as a calculator. The CD-ROM also makes use of ArcView (ArcView version 1, available for free) demonstrations using a media player (Lotus

Figure 1.2 Ordnance Survey of Northern Ireland GIS CD-ROM (Source: reproduced from OSNI CD-ROM by permission of the Ordnance Survey of Northern Ireland on behalf of the Controller of Her Majesty's Stationery Office © Crown Copyright, 2000. Permit number 1519).

Cam) which 'plays' through a series of examples using the ArcView package. As a teaching tool this CD-ROM is very innovative and most useful because it not only offers simple text with illustrations, which can be viewed in a point-and-click Toolbook environment, but also a 'guided tour' through the use of ArcView for some example applications. The development of this resource and its value to the student is clearly highlighted by Galloway:

> By allowing students to input their own data, create links to statistics, perform thematic mapping and explore other spheres of geography, the CD shows how GIS can not only be integrated into the geography curriculum but also help make geography more exciting and relevant.
>
> (Galloway, 1997: 34)

Undergraduate and postgraduate dissertations

A number of undergraduate and postgraduate dissertations have, in recent years also examined the question of GIS education in secondary school education. An undergraduate dissertation completed by Stephen Harris (1991) in the Department of Geography at the University of Aberdeen considered, given the limited funds available to schools for the purchase of hardware and

Example 1:
Data is divided into spacial and attribute. The largest geographical regions are, in turn, located to the left of the sheet and all the sub-regions in them located to their right. If a region has more than one sub-region of the same level, e.g. towns, then these are located vertically downwards under each other. This is repeated for both spatial and attribute data.

Spatial Data (Polygons)			Attribute Data		
SCOTLAND x y - - - - - -	LOTHIAN x y - - - - - -	TRANENT x y - - - - - -	SCOTLAND A. ------- B. ------- C. ------- D. -------	LOTHIAN A. ------- B. ------- C. ------- D. -------	TRANENT A. ------- B. ------- C. ------- D. -------
		MUSSELBURGH x y - - - - - -			MUSSELBURGH A. ------- B. ------- C. ------- D. -------
		HADDINGTON x y - - - - - -			HADDINGTON A. ------- B. ------- C. ------- D. -------
	GRAMPIAN x y - - - - - -	ABERDEEN x y - - - - - -		GRAMPIAN A. ------- B. ------- C. ------- D. -------	ABERDEEN A. ------- B. ------- C. ------- D. -------
		BANFF x y - - - - - -			BANFF A. ------- B. ------- C. ------- D. -------
		PETERHEAD x y - - - - - -			PETERHEAD A. ------- B. ------- C. ------- D. -------
ENGLAND x y - - - - - -	NORTH x y - - - - - -	? x y - - - - - -	ENGLAND A. ------- B. ------- C. ------- D. -------	NORTH A. ------- B. ------- C. ------- D. -------	? A. ------- B. ------- C. ------- D. -------

Figure 1.3a Spreadsheet (after Harris, 1991)

software, the use of existing information technology (IT) as a means to (a) teach GIS, and (b) to teach elements of GIS and applications. The examples in this thesis included applications using spreadsheets and graphics software (figures 1.3a, 1.3b, 1.4a and 1.4b). Simple though the applications are, the ability to convey the basic concepts of GIS were determined to be ideal for school-age children.

A postgraduate dissertation completed by Christine Warr (1996) at the University of Salford considers the requirements for teaching GIS at the school level. In a total of eight chapters the basic requirements for GIS education in schools are considered. The primary advantage of introducing GIS in school education is seen as: '... to provide a continuity of education in GIS and to equip the next generation with the tools to make use of spatial data' (Warr, 1996). The thesis examines the influence of higher education, the work carried out to date on GIS in schools, from the BBC's Domesday Project, through ESRI, the OS, AEGIS and the work of the NCGIA. This study, through the circulation of a questionnaire, examined the current curricula and syllabuses,

Example 2:
Data is divided into large geographic areas, i.e. countries. These groups are then divided into point, line and polygon data and within these groups, smaller geographical areas are defined, i.e. local regions. Spatial and attribute data is then located to the right of these groups.

Spatial Data				Attribute Data		
Scotland						
Points						
GRAMPIAN				area	population	no. schools
Towns	1. (x,y)	2. (x,y)				
Villages	1. (x,y)	2. (x,y)				
Airports	1. (x,y)	2. (x,y)				
LOTHIAN				area	population	no. schools
Towns	1. (x,y)	2. (x,y)				
Villages	1. (x,y)	2. (x,y)				
Airports	1. (x,y)	2. (x,y)				
Lines						
GRAMPIAN				length	width	surface
A-roads	1.	2.				
B-roads	1.	2.				
Tracks	1.	2.				
Paths	1.	2.				
LOTHIAN				length	width	surface
A-roads	1.	2.				
B-roads	1.	2.				
Tracks	1.	2.				
Paths	1.	2.				
Areas polygons						
GRAMPIAN	(x,y) (x,y)......			population	no. cars	no. houses
LOTHIAN	(x,y) (x,y)......			population	no. cars	no. houses
FIFE	(x,y) (x,y)......			population	no. cars	no. houses
England						
Points						
..................>						
..................>						

Figure 1.3b Spreadsheet (after Harris, 1991)

the degree of postgraduate teacher training, the materials used in teaching, and learning resources.

Another thesis by Steve Palladino (Palladino, 1994b, see http://www.ncgia.ucsb.edu/~spalladi/thesis/Abstract.html) (an ex-staff member of the NCGIA) provides a thorough coverage of the work undertaken by him whilst at the NCGIA.

The related technologies

Although this book will ultimately concentrate primarily on GIS, it is unavoidable that one should necessarily also consider remote sensing (RS) and cartography, as well as Global Positioning Systems (GPS), all of which are an integral part of the current definition of integrated GIS. Remotely sensed data, whether in the form of an aerial photograph, a digital image, video or a satellite image, are an increasingly important source of data and information (where information is defined as being distinct from data through the addition of meaning via analysis) for input to a GIS database. Cartography, maps, and

File Edit View Insert Format Tools Window Help

Arial   ± 10 ±

C3   "Abb

GIS_SP"1.WKS

	A	B	C	D	E	F	G	H	I
1									
2		Place	Abbrev	X-Axis	Y-Axis	Hsn-Ass%	Loc Ath%	Own Occ	PrR+Oth%
3		Abbey W	Abb	466	785	1	34	58	8
4		Arsenal	Ars	438	788	1	87	9	4
5		Burrage	Bur	440	782	9	47	34	10
6		Eynsham	Eyn	465	794	0.1	72	26	3.3
7		Glyndon	Gly	443	788	5	63	23	9
8		Herbert	Her	434	775	8	44	39	9
9		Lakedale	Lak	451	784	5	22	60	13
10		Nightingl	Nig	435	780	0.1	83	0.1	17.4
11		Plumstea	Plu	454.2	772.4	14	14	60	12
12		St. Marys	StM	431	789	2	93	1	3.2
13		St. Nicho	StN	463	779	5	20	63	13
14		Shrewsbr	Shr	439	770	10	7	75	8
15		Slade	Sla	462	775	2	29	62	8
16		Thamesm	Tha	465	803	0.1	94	5	1.1
17		Woolwich	Woo	430	782	4	55	20	21
18									
19									

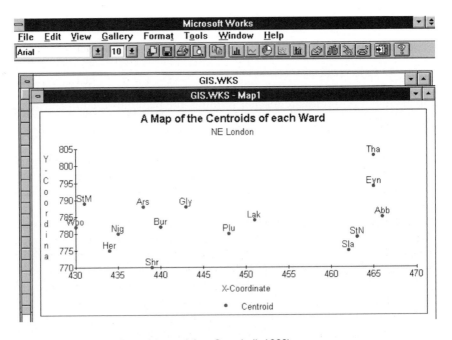

Figures 1.4a and 1.4b Spreadsheets (after Campbell, 1989)

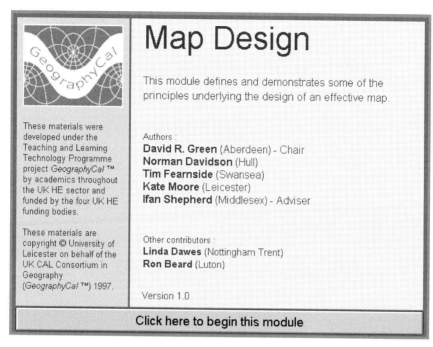

Map Design

This module defines and demonstrates some of the principles underlying the design of an effective map.

These materials were developed under the Teaching and Learning Technology Programme project GeographyCal ™ by academics throughout the UK HE sector and funded by the four UK HE funding bodies.

Authors :
David R. Green (Aberdeen) - Chair
Norman Davidson (Hull)
Tim Fearnside (Swansea)
Kate Moore (Leicester)
Ifan Shepherd (Middlesex) - Adviser

These materials are copyright © University of Leicester on behalf of the UK CAL Consortium in Geography (GeographyCal ™) 1997.

Other contributors :
Linda Dawes (Nottingham Trent)
Ron Beard (Luton)

Version 1.0

Click here to begin this module

Figure 1.5 CTICG T20 Map Design Module

map design are an important part of GIS, most GIS systems being used for map output (good examples of the cartographic output from GIS can be seen in a number of the ESRI publications offering a wide range of examples and uses. The care and attention given to the design of maps is an important component of any GIS education, although in many cases design of a map in an electronic medium is still given scant coverage (see the T20 Map Design Module, figure 1.5).

Global Positioning Systems are also important as they are increasingly used as a means of collecting e.g. survey data which can be uploaded into portable computers or palm tops, even in the field (see Price and Heywood, 1994 and Cornelius et al., 1994).

In the context of remote sensing, the National Remote Sensing Centre (NRSC) (NRSC, 1993) launched a teaching pack 'Satellite Eye Over Europe'. This package was developed jointly with Yorkshire Television, and sought to introduce the use of satellite remote sensing for regional studies. Aimed at Key Stages 3 and 4 of the Geography National Curriculum the package used satellite images for the identification and interpretation of pattern in physical and human geography (written by Keith Orrell, Chief Examiner for GCSE Geography, Midland Examining Group Bristol Project).

It has also been reported that schools in both the UK and France use Meteosat imagery in various school projects (Marchessou, 1995). In Britain a remote sensing option is part of the National Curriculum in schools (Marchessou, 1995: 6). Temple and Vauzelle (1995) described the Dyfed Satellite Project which supports the use of remotely sensed data in primary and secondary education. Whilst materials are available, problems still exist relating to access to suitable information. The AGEN-LLANELLI Remote Sensing Project (1994) had the following objective:

- To give students of the two towns (Agen and Llanelli) the opportunity to learn about the other town and its region
- To make students more aware, by comparison, of the environment of their own region
- To encourage students of Agen and Llanelli to work together
- To develop in the students communication skills appropriate to an international working environment
- To give students the opportunity to use the information technology within an international working environment
- To produce informative details which can be used by other students

This particular project focused on geographical themes, using weather satellite data and multi-spectral SPOT (XS) imagery. Using the TITUS image-processing software (written in France and aimed at the 16–18-year-old age group), students develop hypotheses, which are tested in the field, and process images to identify land-cover information. Caillon (1995) reports on work designed to study satellite images with the aim of understanding geographic, human and physical characteristics of terrain. The work combines maps, satellite images, and field work and computer images.

One part of the work undertaken by the Norwegian Space Centre involves provision of the resources for schools justified in part by the statement that '... Today's pupils are tomorrow's decision takers and it is important to give knowledge about the usefulness of space activities' (Torbo and Stromsholm, 1995: 21). As part of their programme – Satellite Data in Schools – the Norwegian Space Centre aims to:

- concentrate the educational resources to space areas with high national priority
- introduce earth observation in schools
- develop close co-operation with schools

'Svalbard as Seen from Satellites' was a pilot project involving seven primary and secondary schools throughout Norway. Distributed on diskettes, NOAA AVHRR and ERS-1 SAR data, processed by the Tromso satellite station, are part of a package combined with booklets (e.g. 'Earth Observation by Satellites', 'Environmental Monitoring by Satellites') available to school pupils up to the age of 16 (primary level in Norway) offering a chance to study ice situations, glaciers and sea surface characteristics. A disadvantage recognised by Torbo and Stromsholm (1995) is that distribution of images on diskette means that current environmental issues/events cannot be studied. To overcome this disadvantage it is intended that the images will eventually become available over the Internet. This has also been explored by Temple (1995) and the project combined video conferencing, data sharing, and ISDN lines to link schools directly with satellite data from the Satellite Centre. Eight pilot schools were involved, schools who had the Dartcom Weather Satellite Systems provided under the Welsh Office 'Satellites in Schools' initiative. Images could be used on-line with the aid of a tutor.

The value of learning through the use of satellite images has been explored by Roseeu (1995) in the context of 'experience', as a geographical aid, and for transferring information to pupils. The paper discusses some of the practical difficulties associated with delivering, e.g. satellite images to school-age pupils and teachers. It also makes mention of a project conducted with some forty schools in Germany conducted by GEOSPACE GmbH and promoted by DARA. The paper concludes that for effective school education, outside help is definitely needed.

Although not specifically aimed at schools, the UNESCO BILKO image-processing suite of software has potential application within schools. Initially based around a simple suite of DOS-based menu-driven software with a tutorial and some five booklets of image-processing applications to the marine and coastal environment, this software has potential in a secondary school provided it is supported by assistance from teachers. The tutorials are very comprehensive but tend to expect a lot from students. The DOS software has since been replaced by a Windows-based suite of programmes. The software is available from UNESCO (Troost, 1989) and can be downloaded from the BILKO website at http://www.ncl.ac.uk/tcmweb/bilko.

A number of years ago, the International Institute for Aerial Survey (ITC) in the Netherlands developed a suite of integrated remote sensing and GIS software for the Acorn Archimedes series of microcomputers and the Acorn R260 Workstations known as Alexander. This combined remote-sensing digital image-processing software with GIS in a Windows-based environment. Subsequently the software was rewritten for a MS-Windows 3.1 operating system.

Most recently, a number of Internet websites have provided remote sensing materials and tutorials, some of which could, again with appropriate guidance from teachers, offer school pupils insight into the use and analysis of remotely sensed data, sometimes within specific software (e.g. Eric Lorup's remote-sensing tutorial for IDRISI). More recently, a suite of Internet-based remote-sensing tutorials have been developed for higher education by a consortium of universities (e.g. University of Paisley and Edinburgh) (figure 1.6). Mention was made on the NASA Public Access Remote Sensing Internet Projects site about a project designed to 'Enhance the Teaching of Science of Elementary Education through the Application of NASA Remote Sensing Databases and

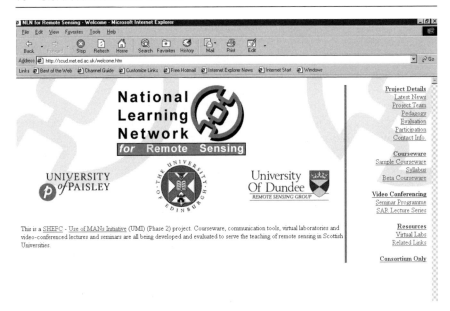

Figure 1.6 National Learning Network for Remote Sensing Tutorial s

Internet Technology', and 'Aquatic Applications of Satellite Imagery' at the K-12 level in the US. In July 1999 EURISY is organising Round Table – 'Integration of Earth Observation into Secondary Education', and includes presentations of earlier seminars held for teachers (EURISY, 1999).

Datasets

It is impossible to ignore the need for data sources in the world of GIS. They are needed for projects and to provide examples. Datasets, whether they are maps, aerial photographs, or satellite images are not always readily available, even if the requirement is for educational purposes. One of the reasons for this is largely cost. Whilst many people would like to have access to data at no cost, good datasets, especially in the UK, usually, and perhaps unfortunately, come at a price even for education. The problem, whether you are in secondary or higher education, is finding the money required to purchase datasets. This is a recurring problem all round, and has been present for many years. In some cases, LAs and some commercial suppliers have made small datasets available to education for free or for a small charge. In fact some datasets come on demonstration diskettes providing sample data e.g. example black-and-white and colour Ordnance Survey raster map data can be obtained from the OS or other commercial suppliers.

In other cases, aerial photography and satellite imagery has been made available for specific projects. But this is not generally the case, and many datasets that could probably be useful to educational institutions around the world still remain out of reach simply because of the cost.

Attempts have been made over the years to try and help UK educational institutions, and most recently help has been offered by the Ordnance Survey and even NRSC Ltd (National Remote Sensing Centre).

Although the demand for datasets in primary and secondary education is probably not as great as in higher education, real datasets are naturally very important for demonstration projects and applications which allow teachers and pupils to work properly with a GIS. True, it is possible to collect one's own data and this in itself can be a useful exercise if it is possible to input data, store, edit, and analyse it, but there are cases where ready-made data is almost a must for educational projects.

GRID-Arendal in Norway has made available a large number of environmental datasets specifically for use in the GIS software IDRISI. These datasets can be downloaded from the GRID-Arendal website at http://www.grida.no/gis/index.htm.

Providing images and maps are available, together with a digitiser or scanner, then it is also possible to create new datasets yourself, e.g. on-screen digitising in ArcView using a scanned map (.TIF format) as a backdrop. Observing copyright of source documents must, however, be seriously considered and must be taken into account by all concerned. Breach of copyright is a serious offence and pupils must not be exposed to such activities or encouraged to treat such actions as the norm. Sometimes it is possible to acquire new spatial map data using GPS, for example. Alternatively if an aerial photograph can be scanned, then it is possible using some of the on-screen digitising tools now available (e.g. in ArcView), to digitise real world features as points, lines and areas (polygons). If possible this makes it possible for pupils to start to use the data capture functionality of GIS as well (figure 1.7).

Information technology

Last but not least it is important to consider information technology (IT) in general (hardware and peripherals), as well as the use and application of software, e.g. spreadsheets, databases, graphics software, statistical packages and so on, in the context of GIS. Whilst many GIS now include similar functionality to the above software within the system, other GIS functionality relies upon additional software – whereby the data is exported to e.g. a statistical package for analysis, and the results are re-imported back into the GIS. It is also possible to use basic desktop IT as the basis with which to teach elements of GIS in school education as demonstrated by Harris (1991) and Warr (1996). A simple approach was in fact used in some research work by Coggins and Brown (1992) who made use of various different software as so-called functional modules of a GIS. The system developed as part of a study into household waste recycling, used an Apple Macintosh computer.

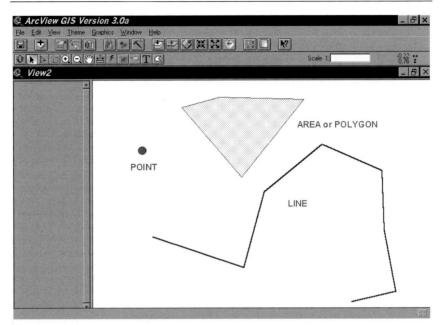

Figure 1.7 ArcView Digitising (Graphic image supplied courtesy of Environmental Systems Research, Inc. and is used herein with permission)

By and large, whatever software is purchased today it will usually be designed to run on a personal computer (PC) or an Apple Macintosh micro-computer platform, under MS-Windows (3.1, 95, 98 or NT [New Technology]) or the Apple OS. The more sophisticated the software and the OS, then probably the more powerful the hardware (processor, RAM, graphics and modem speed) will need to be. Although it is still possible to find GIS software that will run on a basic PC under DOS (e.g. the older versions of IDRISI, PMAP and AMAP), most PCs purchased now will probably be running W95/98/2000/NT.

Purchasing a basic computer system is not really a viable solution for any GIS or digital image-processing software, and the message that should be put out to school, IT purchasers now is 'buy the best and most powerful computer hardware that you can for the money'. As the technology continues to develop, computer systems are very rapidly becoming out of date in a short period of time. It must also be remembered that new versions of software usually come with new functionality, and more often than not they consume even greater resources than the last version in terms of the hard disk space required to store the OS and the software, demands on the display devices, e.g. colour monitors, and so on. Whatever GIS software is selected for use, then unless it is very basic, it will demand adequate resources e.g. ArcView. Furthermore, plenty of hard disk space will be required for storing raster data e.g. a single scanned colour aerial photograph of 9in × 9in (23 cm × 23 cm) at 300 dpi and 24-bit colour will require some 22Mb ((([dimension × dpi] × [dimension

× dpi] × 24)/8/1,000,000). Similar demands will be made on the requirements for display memory. Whilst such demands will not be relevant perhaps for teaching and using GIS in a primary school, it may be an important requirement to have an upper end specification system for practical applications in a secondary school. Similarly, access to the Internet requires plenty of RAM, and a fast modem, or ideally an ISDN line. Computer platform specifications for ArcView, for example, are shown in table 1.1.

Summary and conclusions

In writing this book, there has been a conscious attempt to try and provide a logical structure to the order of the chapters included. In other words a 'start and finish' to the teaching of GIS in education – not just schools but also in further and higher education. The purpose and value of providing the links between secondary and higher education is to emphasise the need for continuity in the education system. Whilst not all countries teach geography in the school or higher-education curriculum, GIS does undoubtedly provide a strong focus for geography as a subject, whilst also pulling together elements of other subjects in the curriculum (e.g. mathematics and statistics) that are relevant to geography. Although adoption of GIS has not been universal within the UK school system, in part because of the technical/technological nature of the subject and I suspect a reluctance and fear by many teachers to become involved in the application of information technology, the ESRI software packages and examples clearly indicate how important GIS is to society. This is perhaps strengthened by the growing number of maps now being posted and used on the Internet, and the tremendous growth in Internet-based GIS. Today there are some exceptional examples of the use of the Internet as a means to deliver maps, either static or dynamic, to a large number of people. These include simple image maps, maps as .PDF files (Adobe® Acrobat® portable document files), Internet GIS-served maps and also maps delivered 'on the fly'.

GIS is, together with the related technologies of remote sensing, digital mapping (cartography), GPS, and most recently the Internet, a tremendous new development. Its role in everyday life is something that people of all

Table 1.1 Computer platform specifications for ArcView 3.1.

ArcView GIS Version 3.1 runs on

– Microsoft® Windows® (95/98, NT)
– Digital™ UNIX® Alpha™
– HP® 9000/700 and 8x7
– IBM® RS/6000™
– Silicon Graphics®
– Solaris® 2

---

ages should be made more aware of. Whilst a very technical and complex subject area in itself, the visual side of GIS – the map output (one form of visualisation) – is something that can be used to attract people's attention. This visual aspect is most clearly seen in many of the ESRI publications, and by browsing on the Internet.

References

AGI (1997). The Association for Geographic Information: AGIS – GIS for Schools. Available HTTP at: http://www.geo.ed.ac.uk/agi/schools/agis.html

Arnberg, W. (1995). 'GIS in School – The National PC-Atlas of Sweden', *Remote Sensing*, Information from the Swedish Space Corporation, No. 27, December, pp. 20–4.

AUME (1990). Press Release 1: NCET/DES National Curriculum Software Scheme: Geographical Information System for Schools, 2p.

Autodesk (1997a). Kingston, Ontario – Canada's Digital City. Available HTTP at: http://www.autodesk.com/solution/edu/ebd/winter97/ebdw9703.htm

Autodesk (1997b). Education by Design: Pilot Schools Stake their Claims to Community-Based GIS. Available HTTP at: http://www.autodesk.com/solution/edu/ebd/winter97/ebdw9702.htm

Cadoux-Hudson, J. and Heywood, D.I. (1991). Eds, *Geographic Information 1991. The Yearbook of the Association for Geographic Information*. Taylor & Francis/Miles Arnold, London, 421p.

Cadoux-Hudson, J. and Heywood, D.I. (1993). Eds, *The Yearbook of the Association for Geographic Education: Geographic Information 1992–1993*, Taylor & Francis, London.

Caillon, M. (1995). 'Atlantic Ark – Producing Audio-Visual Material on the Environment for Young People in Two European Countries', in Proceedings of a EURISY/Norwegian Space Centre Workshop: Earth Observation from Space as a Resource for Teaching, Andoya, Norway, June 15–16, pp. 16–20.

Campbell, N. (1989). 'The Spreadsheet as a Low Cost Geographical Information System', Paper presented at a Workshop on the New Technology and Spatial Information Systems, University of Nottingham, January, 16p.

Cassettari, S. (1991). 'Introducing GIS into the National Curriculum for Geography, GIS in the Secondary and Higher Education Curriculum', Proceedings of AGI'91, November 20–22, ICC, Birmingham, pp. 3.4.1–3.4.4.

Chaloner, M. (1992a). 'GIS Education in the News!', *Mapping Awareness and GIS in Europe*, Vol. 6(5), June.

Chaloner, M. (1992b). 'GIS in European Schools, Emulation, Education, Geography and GIS', *Mapping Awareness and GIS in Europe*, Vol. 6(6), July/August, pp. 30–1.

Chaloner, M. (1992c). 'Take Away GIS! Multimedia/Hypermedia as a Teaching Resource in GIS Education', *Mapping Awareness and GIS in Europe*, Vol. 6(7), September, pp. 46–7.

Chaloner, M. (1992d). 'Education and Training is Crucial to Our Future', *Mapping Awareness and GIS in Europe*, Vol. 6(8), November, pp. 45–6.

Clark, M.J. (1991). 'GIS Awareness: The Technological and Educational Challenge, GIS in the Secondary and Higher Education Curriculum', Proceedings of AGI '91, November 20–22, ICC, Birmingham, pp. 3.9.1–3.9.4.

Coggins, C. and Brown, R. (1992). 'GIS Monitors Household Waste Recycling', *GIS Europe*, Vol. 1(3), June, pp. 34–5.

Cornelius, S., Carver, S., Heywood, I. and Sear, D. (1994). 'Mountains and Mosquitoes. Using GIS in the Field', *GIS Europe*, Vol. 3(1), February, pp. 42–4.

ESRI (1993). 'Touching the Future: Students Show Interest in GIS. Desktop News', *ARC News*, Spring, p. 19.

ESRI (1997). Schools and Libraries Program: GIS for Schools, Libraries, and Other Neat Places. Available HTTP at: http://www.esri.com/base/markets/k-12/k-12.html

ESRI, (1998a). *GIS in K-12 Education*, an ESRI White Paper, March, 36p.

ESRI, (1998b). 'GIS Touches All Our Lives, Everyday', *ESRI ARC News*, Summer, p. 34.

EURISY (1999). *EURISY Newsletter*, No. 29, January.

Fitzpatrick, C. (1993). 'ArcSchool Reader Makes Debut', *Arc News*, Summer, p. 38.

Forer, P. (1994). 'The Python and the Pig: Absorbing GIS into Everyday Education', in *Geographic Information 1994 The Sourcebook*, (Eds D.R. Green, D. Rix and J. Cadoux-Hudson) pp. 355–63, Taylor & Francis, London.

Freeman, D. (1991). 'Development of an Educational GIS for the National Curriculum, GIS in the Secondary and Higher Education Curriculum', Proceedings of AGI '91, November 20–22, ICC, Birmingham, pp. 3.6.1–3.6.4.

Freeman, D. (1993). 'GIS in Schools'. Available HTTP at: http://www.agi.org.uk/pages/freepubs/gisinschoolsabs.html

Freeman, D. (1993). 'GIS in Schools', *AGI Publication* No. 4/93, October, Information and Education Committee Publication, Association for Geographic Information: London.

Freeman, D., Green, D., Hassell, D. and Paterson, K. (1993). 'Getting Started with GIS', *Teaching Geography*, April, pp. 57–60.

Freeman, D., Green, D. and Hassell, D. (1994). 'A Guide to Geographic Information Systems (GIS)', *Teaching Geography*, January, pp. 36–7 (Information Technology Section).

Galloway, B. (1996). 'GIS in the Northern Ireland Schools Curriculum – A Multimedia Approach!'. Proceedings of AGI GIS'96.

Galloway, B. (1997). 'Schools of Thought – Putting GIS on the Curriculum', *Mapping Awareness*, Vol. 11(3), pp. 32–4.

Gill, S. and Roberts, P. (1998). 'GIS in Every High School in Powys, Ordnance Survey Education. Available HTTP at: http://www.o-s.co.uk/educate/mapnews/powys.htm

Gittings, B.M., Healy, R.G. and Stuart, N. (1993). 'Towards the Twenty-First Century in GIS Education', in *The Yearbook of the Association for Geographic Education: Geographic Information 1992–1993* (Eds J. Cadoux-Hudson and D.I. Heywood) pp. 326–34, Taylor & Francis, London.

Green, D.R. (1991a). 'Making Space: Including Education Sessions at Commercial GIS Conferences', *Mapping Awareness*, Vol. 5(8), October, pp. 40–2.

Green, D.R. (1991b). 'GIS Software for Schools: What About It?', *Mapping Awareness*, Vol. 5(9), November, pp. 34–6.

Green, D.R. (1992a). 'A Mixed Bag this Month', *Mapping Awareness and GIS in Europe*, Vol. 6(3), May, pp. 30–1.

Green, D.R. (1992b). 'GIS in Secondary Education: Recent Developments in the US and UK', *Mapping Awareness and GIS in Europe*, Vol. 6(4), May, pp.72–3.

Green, D.R. (1993). 'GIS Education and Training: Developing Educational Progression and Continuity for the Future', in *The Yearbook of the Association for Geographic Education: Geographic Information 1992–1993* (Eds J. Cadoux-Hudson and D.I. Heywood) pp. 283–95, Taylor & Francis, London.

Green, D.R. (1994). 'Introduction: GIS in Education and Training – Progress in 1993', in *Geographic Information 1994 The Sourcebook* (Eds D.R. Green, D. Rix and J. Cadoux-Hudson), pp. 329–31, Taylor & Francis, London.

Green, D.R. and McEwen, L.J. (1991). 'GIS as a Component of IT Courses in Higher Education Geography Courses', in *The Association for Geographic Information Yearbook 1990* (Eds M.J. Foster and P.J. Shand) pp. 287–94.

Green, D.R. and McEwen, L.J. (1992). 'Turning Data into Information: Assessing GIS User Interfaces', Proceedings of AGI'92, November 24–26, ICC, Birmingham, pp. 1.27.1–1.27.10.

Green, D.R. and McEwen, L.J. (1994). 'The User-Friendly Interface: An Essential for GIS Education and Training', in *Geographic Information 1994 The Sourcebook* (Eds D.R. Green, D. Rix and J. Cadoux-Hudson), pp. 355–63, Taylor & Francis/Miles Arnold, London.

Green, D.R., Rix, D. and Cadoux-Hudson, J. (1994). Eds, *Geographic Information 1994 The Sourcebook*, Taylor & Francis, London.

Harris, S. (1991). 'Emulating GIS in Education', Unpublished BSc Thesis, University of Aberdeen.

Heywood, I. and Petch, J.R. (1991). 'GIS: A Toybox Approach, GIS in the Secondary and Higher Education Curriculum', Proceedings of AGI'91, November 20–22, ICC, Birmingham, pp. 3.5.1–3.5.3.

IDRISI (1996). 'IDRISI Use Expands in Schools', *IDRISI News*, Fall, Vol. 8(1), p.15.

IDRISI (1998). IDRISI in Schools. Available HTTP at: http://www.idrisi.clarku.edu/07school/main.htm

Kemp, K.K. (1991). 'GIS Education Around the World: Year 3 of the NCGIA Core Curriculum Project, GIS in the Secondary and Higher Education Curriculum', Proceedings of AGI'91, November 20–22, ICC, Birmingham, pp. 3.8.1–3.8.3

Langford, M. and Strachan, A.J. (1991). 'Getting Started in GIS: Cost Effective Solutions and Academic Support, GIS in the Secondary and Higher Education Curriculum', Proceedings of AGI'91, November 20–22, ICC, Birmingham, pp. 3.7.1–3.7.4

Maguire, D.J. (1993). 'Teaching Earth and Social Science in Higher Education using Computers: Bad Practice, Poor Prospects?', in *The Yearbook of the Association for Geographic Education: Geographic Information 1992–1993* (Eds J. Cadoux-Hudson and D.I. Heywood) pp. 313–17, Taylor & Francis, London.

Marchessou, F. (1995). 'The Educational Offer: A European Overview', in Proceedings of a EURISY/Norwegian Space Centre Workshop: Earth Observation from Space as a Resource for Teaching, Andoya, Norway, June 15–16, 1995, pp. 4–9.

Masser, I. and Toppen, F. (1994). 'The EGIS European GIS Education Special Interest Group', in *Geographic Information 1994 The Sourcebook*, (Eds D.R. Green, D. Rix and J. Cadoux-Hudson) pp. 370–4, Taylor & Francis, London.

NCGIA (1993a). 'GIS in the Schools', *NCGIA Newsletter*, May, 2p.

NCGIA (1993b). 'GIS in the Schools', *NCGIA Newsletter*, November, 4p.

NCGIA (1999). NCGIA Secondary Education Project Materials. Available HTTP at: http://www.ncgis.ucsb.edu/education/projects/SEP/seppubs.html

Newell, R. (1997). 'GIS in Schools', *Geographic Information – The Newsletter of The Association for Geographic Information*, Vol. 7(1), January, p. 12.

NRSC (1993). 'Satellites Bring Europe into the Classroom', *News Release*, Tuesday January 5, 1p.

Ordnance Survey (1999) 'Ordnance Survey: Geographic Information System (GIS)'. Available HTTP at: http://www.doeni.gov.uk/ordnance/gised.htm

Palladino, S.D. (1993a). 'GIS in the Schools: Workshop Resource Packet', *Technical Report 93-2*, Santa Barbara, California, National Center for Geographic Information and Analysis.

Palladino, S. (1993b). 'GIS and Secondary Education in the US', in *The Yearbook of the Association for Geographic Education: Geographic Information 1992–1993* (Eds J. Cadoux-Hudson and D.I. Heywood) pp. 304–9, Taylor & Francis, London.

Palladino, S. (1994a). 'An Update on GIS in the US Secondary Schools and NCGIA Secondary Education Project Activities', in *Geographic Information 1994 The Sourcebook*, (Eds D.R. Green, D. Rix and J. Cadoux-Hudson) pp. 340–5, Taylor & Francis, London.

Palladino, S. (1994b). 'A Role for Geographic Information Systems in the Secondary Schools: An Assessment of the Current Status and Future Possibilities, Unpublished MSc Thesis, University of California, Santa Barbara.

Palladino, S. and Goodchild, M.F. (1993). 'A Place for GIS in the Secondary Schools? Lessons from the NCGIA Secondary Education Project', *Geo Info Systems*, April.

Palladino, S. and Kemp, K.K. (1991). 'GIS Teaching Facilities: Six Case Studies on the Acquisition and Management of Laboratories', *Technical Report 91-21*, Santa Barbara, California, National Center for Geographic Information and Analysis.

Palladino, S. and Van Zuyle, P. (1996). 'Critical Issues in GIS-Based Educational Module Development: NCGIA's ArcView-based Color Your World Module', *Technical Report 96-6*, Santa Barbara, California, National Center for Geographic Information and Analysis.

Petch, J.R., Reeve, D., Cornelius, S.C., Heywood, D.I. and Burgess, D. (1994). 'Postgraduate Distance Learning in GIS', in *Geographic Information 1994 The Sourcebook*, (Eds D.R. Green, D. Rix and J. Cadoux-Hudson) pp. 364–9, Taylor & Francis, London.

Price, M.F and Heywood, D.I. (1994). Eds, *Mountain Environments and Geographic Information Systems*, Taylor & Francis, London.

Raper, J. (1993). 'GIS Teaching without a GIS', in *The Yearbook of the Association for Geographic Education: Geographic Information 1992–1993*, (Eds J. Cadoux-Hudson and D.I. Heywood) pp. 318–25, Taylor & Francis, London.

Rhind, D. (1993a). 'Maps, Information and Geography: A New Relationship', *Geography*, Vol. 78(2), No. 339, pp. 150–9.

Rhind, D. (1993b). 'Ordnance Survey and Schools', *Teaching Geography*, Vol. 18(1), pp. 18–19.

Roseeu, R. (1995). 'Educational Background', in Proceedings of a EURISY/Norwegian Space Centre Workshop: Earth Observation from Space as a Resource for Teaching, Andoya, Norway, June 15–16, pp. 26–33.

Sharpe, B. and Crechiolo, A. (1999). 'Teaching Geography with GIS in Ontario's Secondary Schools'. Available HTTP at: http://ncgia.ucsb.edu/conf/gishe96/program/sharpe.html

Smith, G. (1994). 'GIS and Secondary Education in OS', in *Geographic Information 1994 The Sourcebook*, (Eds D.R. Green, D. Rix and J. Cadoux-Hudson) pp. 336–9, Taylor & Francis, London.

SPANS® (1995). 'TYDAC Plays Important Role in GIS Education Project', *The SPANS Dialogue Box*, February, p. 3.

Temple, A. (1995). 'The Superhighway: Opening the Door to Satellite Remote Sensing', in Proceedings of a EURISY/Norwegian Space Centre Workshop: Earth Observation from Space as a Resource for Teaching, Andoya, Norway, June 15–16, pp. 34–7.

Temple, A. and Vauzelle, M. (1995). 'The Need for Earth Observation in Primary and Secondary Schools: Forging Links Across Europe', in Proceedings of a EURISY/Norwegian Space Centre Workshop: Earth Observation from Space as a Resource for Teaching, Andoya, Norway, June 15–16, pp. 10–15.

Torbo, P. and Stromsholm, B. (1995). 'Svalbard as Seen from Satellites', in Proceedings of a EURISY/Norwegian Space Centre Workshop: Earth Observation from Space as a Resource for Teaching, Andoya, Norway, June 15–16, pp. 21–5.

Troost, D. (1989). 'Some Marine Applications of Satellite and Airborne Remote Sensing', MarinF/70, UNESCO TREDMAR, July 1989.

Unwin, D. and Dale, P. (1991). 'An Educationalist's View of GIS', in *The Association for Geographic Information Yearbook 1990* (Eds M.J. Foster and P.J. Shand) pp. 304–12, Miles Arnold, London.

Unwin, D. (1990). 'A Syllabus for Teaching Geographical Information Systems', *International Journal of Geographical Information Systems*, Vol. 4(4), pp. 457–65.

Vicars, D. (1993). 'GIS and Higher Education', in *The Yearbook of the Association for Geographic Education: Geographic Information 1992–1993* (Eds J. Cadoux-Hudson and D.I. Heywood) pp. 310–12, Taylor & Francis, London.

Warr, C.M. (1996). 'GIS in Secondary Education in England and Wales: From Vision to Reality?', Unpublished MSc Thesis in Geographical Information Systems, Salford University.

Wheatley, R. (1994) 'Langdon Park School', in *Geographic Information 1994 The Sourcebook*, (Eds D.R. Green, D. Rix and J. Cadoux-Hudson) pp. 332–5, Taylor & Francis, London.

Wood, A. and Cassettari, S. (1993). 'GIS and Secondary Education', in *The Yearbook of the Association for Geographic Education: Geographic Information 1992–1993* (Eds J. Cadoux-Hudson and D.I. Heywood) pp. 296–303, Taylor & Francis, London.

Further reading and information

Carvel, J. (1998). 'Computer Revolution will Connect Schools to a National Grid for Learning, Preventing Emergence of an "Information Poor" Generation: Blair's £1bn IT Package for Schools'. *The Guardian*, Saturday November 7, Home News, p. 7.

EURISY (1995). *Earth Observation from Space as a Resource for Teaching*, Workshop, Andoya (Norway) June 15–16, Norwegian Space Centre, October 1995, 77p.

GIS in Education: Wisconsin Watershed Watch: GIS in Education for Ecosystem Health (1999). http://danenet.wcip.org/gisedu/about.html

National Council for Geographic Education (1999). *National Council for Geographic Education*. Available HTTP at: http://snyoneab.oneonta.edu/~baumanpr/ncge/rstf.htm

National Curriculum (1999). *The National Curriculum – Geography Key Stage 1,2,3.* Available HTTP at: http://www.dfee.gov.uk/nc/geoks1.html, http://www.dfee.gov.uk/nc/geoks2.html and http://www.dfee.gov.uk/nc/geoks3.html

NCET (1997). *Geography and the Internet.* Available HTTP at: http://easyweb.easynet.co.uk/~rallen/NCET/

NCET (1991). *I.T.'s Geography: The Role of I.T. in P.G.C.E. Geography Courses*, NCET Occasional Paper 3, information technology in Initial Teacher Training, 81p, plus Appendix (5p).

NCGE (1999). Geography for Life: The National Geography Standards. Available HTTP at: http://www.ncge.org/tutorial/index.html, http://www.ncge.org/publications/teacher/, http://www.ncge.org/publications/importance/, http://www.ncge.org/publications/family/ and http://www.ncge.org/publications/business/

Chapter 2

Geographical information in schools

Past, present and future

Jim Page, Gwil Williams and David Rhind

A short history of geography

At least one of these authors is old enough to remember a very different subject being taught in schools! Forty years ago geographical information in schools reflected the reactive and encyclopaedic nature of the subject. The world situation had an air of permanence: rubber came from Malaysia, grain came from America, coffee came from South America. A large fraction of our exports went to the Commonwealth nations but in many other nations we were a major supplier of manufactured goods. At home, the Midlands were the manufacturing heartland of the UK, the North West, North East and Scotland built ships for a thriving British fleet, and South Wales was a major coal mining centre.

Mapping reflected this apparent permanence and defined the detail of country boundaries, major rivers, the location of principal towns and cities, the types and location of manufacturing and agriculture. In the classroom, care was still taken to ensure that such knowledge was learnt by rote and only exceptionally were concepts and processes of analysing the reasons behind the 'facts' considered relevant. Everyday matters of modern geography – such as the environment and pollution – were just not significant issues at that time. Physical geography still consisted of the study of rivers from source to sea – Davis's three ages – coastline formations, types of rainfall, erosion processes, types of landscape and their identifying features with the need for detailed knowledge of terms – oxbow lakes, scarp slopes – writ large. The idea of visiting the field and experiencing geography 'in the wild' in most cases did not apply. Geography was primarily inculcated via 'talk and chalk'. Some skills were highly developed: typically, geography teachers could draw a perfect circle on the board every time when illustrating world issues! It was a black-and-white world, the only colours normally in use being the pastel shades of chalk and roneo inks. Resources usually consisted of one principal text (like Unstead or Briault and Hubbard), world and British Isles wall maps on linen, a world atlas and (of course) Ordnance Survey map extracts for map interpretation and national grid referencing exercises.

The purpose of this description of pre-history is not to expose our ante-diluvian pasts. Nor in any way is it to demean those responsible for teaching the subject over three decades ago. On the contrary, all three of us became what we became because of inspiring teachers. The purpose is rather to remind the reader of just how much has changed in the way we look at the world and, in particular, how geography has evolved since then.

The development of practical local fieldwork exercises and investigative field visits as part of geography teaching during the 1960s (and ever since) began to move the subject from a rigid, reactive, taught-by-rote one to a proactive, analytical and live entity. Pupils exposed to 'raw geography' soon became aware that reality did not always fit the neat textbook model. Discussions with people in the field revealed conflicts of interest and differences that did not equate to a neat boundary on a map or text definition. National and world economics, politics and social considerations came ever more into play. Enquiry-based learning was born. Teachers, freed from the classroom routine and empowered by the advent of the photocopier, produced their own resource material including games, role plays and simulations to develop the enquiring and analytical approach. Information on world issues was also becoming more readily available and the simple, single-source model of long-lasting books as major information sources began to decay. Schools' programmes and current affairs broadcasts on radio and TV also provided a new dimension, supplementing and developing awareness far beyond the textbook example. Thus, while classrooms still contained the definitive text, world atlas and linen wall maps, new key supply channels of knowledge and arguments came into use – newspaper articles, fact sheets, field work reports, study notes, audio tapes and television. Colour entered the description of the world as well as its reality. The information age in geography had begun.

Our present world

The upside of the information revolution

Enquiry-based learning, and analysis and interpretation of information to reach conclusions, is re-enforced by the new National Curriculum for geography introduced in September 1995. Studies that focus on geographical questions such as 'how and why is it changing?', 'what are the implications?' and 'how did it get like this?' join the more prescriptive 'what is it?', 'where is it?' and 'what is it like?'.

The geography curriculum, as with that for all subjects, stipulates the use of information technology (IT) to gain access to additional information sources and to assist in handling, presenting and analysing geographical evidence. The introduction of computers and the links they facilitate into worldwide information sources underpins a revolution in geography teaching and learning. In this respect, Britain is very fortunate: though computing facilities in many schools remain inadequate and parents now have to play a major role in supporting IT in schools, there are more computers per pupil than almost anywhere else in the world.

In principle, these IT tools also make possible the collecting, collating and analysing of a variety of information sets linked to a geographical area or location. Nor is this something particular to education: the same processes are being applied every day in business using Geographical Information Systems (GIS). The GIS industry worldwide is now reckoned to be worth around £5 billion annually. Applications of Geographical Information Systems range from recording assets (such as pipelines or networks of the utility companies and land registration details), through optimising locations (for sites of supermarkets or the routing of garbage trucks) to analysing environmental problems at the local or global scales. A vital factor is that such GIS facilitate the linking of information databases via geographical frameworks such as Ordnance Survey digital map data. At school level, this can mean recording ages and types of buildings in the locality and then plotting the information on top of OS data to create a picture of development growth in the area. At business level, the same processes can involve an insurance company analysing the risk of subsidence in an area. By plotting the ages of buildings on an Ordnance Survey framework with a backdrop of geological information and the postcode and address of each property, the company can then use the analytical power of the GIS to establish which properties are most at risk, thereby being able to target premiums more precisely and competitively.

The same computer, with relevant software and OS Digital Elevation Model data (a grid of heights across the ground surface) also facilitates the teaching of physical geography, providing the capability to display mapping as a three-dimensional (3-D) model and to view it from any angle and elevation. Software to do this was recently given away with one *Personal Computer* magazine! No more is there the need to teach the interpretation of a 3-D landscape from a two-dimensional (2-D) map and spend hours constructing models out of polystyrene layers cut to contour lines or the drawing of cross sections to study intervisibility. These tools are by no means limited in their application to geography: the terrain model can be used just as readily in mathematics as a readily understandable representation of three-dimensional co-ordinates. Thus OS data and the software tools are a truly cross-curricular resource.

There are obvious issues of suitability, relevance and cost to address but systems such as British Telecom's (BT's) Campus World, to which a variety of major organisations (including Ordnance Survey) contribute, provide a 'walled garden' of educationally relevant material with access to the wider World Wide Web if required. So Horndean School's geography department (at which one of us teaches), like many others, has now grown its information resources to include access to a computing facility with Ordnance Survey digital map data, relevant software packages for mapping and access to the World Wide Web.

Thus far, we have dealt with a very British perspective on information sources. But what of worldwide issues? The Internet and World Wide Web provide global access to 30 million computers, many of which hold some geographical information – if you know how to find it. For example, within days of the Kobe earthquake in Japan, one of the schools in the city had established a Web page recording their experiences and how they were coping in the aftermath. This was updated every day and was accessible from any one of the 30 million or more computers linked to the Internet. Alongside this, the Geographical Survey Institute in Japan – the equivalent of OS – made available frequently updated information on the size, nature and frequency of the shocks and what officials were doing to minimise the problems. All this information is free, as is much that is held in the greatest libraries in the world (at least those in the USA). Breathtakingly powerful search engines can be unleashed to seek out and retrieve information meeting a combination of criteria. Population statistics, the weather almost anywhere in the world, satellite and shuttle photographs of earth are accessible to all – at least in theory.

One specific case where education has exploited high technology on many fronts is in the state of North Carolina. The state government has seen the development of high technology as underpinning the transition of their economy from an industrial/tobacco-based one to an information economy. The use of the latest technologies, allied to strict adherence to standards, has been an explicit policy for over twenty years. The world's first routinely operational high-speed asynchronous transfer mode (ATM) network is now functioning across North Carolina and permits interactive video piped into classrooms and health care sites across the state, irrespective of location and at the same cost (thanks to a tariff-smoothing strategy). By early 1997, the network will be running at nearly 1 gigabits per second (i.e. 1 million million bits). As a result, groups of classes get taught by top experts in certain subjects, these experts being located far away elsewhere in the state. To an extent, this has the effect of removing the geography of resource constraints. On the other hand, the state is a leading one in developing GIS and using it in state-wide decision-making! This reliance on technology is allied to other policies (e.g. setting up a maths and science institute of those teachers who are able to teach effectively over the network). In many respects, the North Carolina approach is a highly *dirigiste* one but the evidence of very low unemployment rates, high skills levels and considerable inward investment suggests that the government has met its objectives.

Based both on the promise of things to come and the existence of some of this in certain favoured places now, the technology should provide an unrivalled and dynamic teaching resource for those with the knowledge, interest and resources to access such information.

The downside

The downside of all this has been articulated by a variety of critics. The most common criticisms have been:

- the social consequences of a technology-push approach are often harmful and can be dehumanising
- some of the information is positively misleading, particularly since quality control for much of it is wholly absent or undefined
- filtering, retrieval and analysis of information is often much more difficult to make work satisfactorily than the enthusiasts acknowledge
- it imposes new costs (and stresses) on those in education
- it creates a culture of secondary data analysts who believe that all problems can be resolved by tapping at a keyboard
- it fosters laziness and lack of analysis in that students often simply 'cut and paste' vast amounts of material from a CD-ROM into their projects

These criticisms have some force. For instance, efficiency gains in setting insurance premiums on houses in relation to the real geological or social hazards may 'red line' some areas and render properties within them uninsurable. Equally, a mindless use of many computer games is scarcely likely to advance citizenship or understanding of the world. Yet the 8 million copies sold of SimCity testify to the educational power of those games which fascinate, challenge and involve children and adults alike in a simulation relevant to real-world concerns. The complexity of dealing with unpredicted events, the interactions between actions taken and the consequences of ill-conceived actions are far better demonstrated by SimCity than by the arid study of static examples.

Some of the criticisms are no less real but are likely to be transitory. For instance, the available cost of bandwidth ensures that schools accessing information sources can often consume telephone time and hence financial resources in short supply elsewhere in the organisation. Equally, what looks simple in a demonstration by a visiting expert can place high demands on the time of and training opportunities for the classroom teacher: there is often a substantial opportunity cost in IT-based geography. But bandwidth is becoming cheaper faster than the growth of information utilised and major efforts are being made by vendors like Microsoft to produce 'plug and play' applications for geographical information as much as in encyclopaedias like *Encarta*. Interestingly, a global homogenisation of knowledge on an American basis, underpinned by US examples, seems to be in the process of being suspended. Major corporations such as Microsoft are recognising the importance of local content in fostering acceptability in highly differentiated markets. Thus whilst structures and concepts may be set by corporate headquarters, the examples may increasingly be locally controlled. In this way, some economies of scale can still be realised. A consequence of all this is the fostering of the English language as a lingua franca.

So far as the quality of information is concerned, there is one safe route to follow: wherever possible, use the offerings of organisations which have long been in existence, have a reputation for quality and which are happy to tell you how their data is collected and what is its specification, including its accuracy. Examples of these organisations in the UK include the Meteorological Office, the Office for National Statistics and (of course!) Ordnance Survey.

Looking forward

What of the future? It is easy to paint a picture where the potential is limitless, fuelled by constantly improving technology and the freedom of access to worldwide information, coupled with an IT-aware population of students. One scenario has it that availability of information will not be a problem. It will simply be a question of identifying relevant material and applying it, together with new technology, to meet the needs of the subject curricula. Changes in the curricula can thus be accommodated more easily than at present because the required information can be sought – and found – at the touch of a button. Specific geographical resources for schools will be increasingly available in electronic form (CD-ROM and online) with virtual landscapes and scenarios that will allow users full interaction with content material. Individual teachers will set up and release their own electronic agents ceaselessly hunting through electronic databases to find information matching a set of criteria relevant to the lessons of, say, Year 10 geography. They will publish their own field area notes on the Internet and access those created by their colleagues for different areas; such publishing is hugely cheaper and easier than traditional paper-based books or pamphlets. Video-conferencing facilities will enable school pupils to see and talk directly to each other and to exchange information, views and ideas on geographical and other issues. In Horndean School, for instance, the geography department has become involved in the Comenius Action 1 programme of the European Commission. Links have been forged with schools in Austria and Belgium using video conferencing.

It is important to recognise that a growth in use of computers will arise not only because they provide access to new information sources but because they will integrate access to *all* information – written text, speeches, images, maps and videos could, in principle, be accessed from a common catalogue (akin to the use of ISBNs), retrieved and 'played' as needed 'on demand', thereby displacing all video recorders, overhead projectors and even photocopiers. This assumes utter reliability on the part of all the hardware, software and telecommunications!

Such a scenario is scarcely credible because it is posited on 'technology push', rather than on any curriculum design or human and institutional preferences and frictions. For instance, existing publishers have little interest in being rendered bankrupt by the supplanting of their books by a set of nerdish teachers surfing on the Internet! Moreover, few teachers are likely to have the time (or even skills) to constantly update their own knowledge base. What seems much more likely to happen is a growth of use of multimedia allied to some traditional teaching resources, the latter being updated in part by authors themselves using the Internet. For example, a set of standard notes and/or handouts might contain key statistics and maps. If the latter were drawn from centralised sources and 'hot-linked' they could be updated annually without further action by the teacher: interpretations of trends made by experts could be switched on or off at will.

The most significant problem of all is who will pay for all this new world order. So far as telecommunications are concerned, this seems likely to be the users. The decrease in costs of telephony and other networks is likely to continue in the foreseeable future and bring many other applications within reach of the 'average school'. But this disguises two other problems. The first is that schools vary greatly in their purchasing power. The ability of an inner-city comprehensive to obtain the necessary resources is always likely to be less than a major grant-maintained school or a major public school. To that degree, there is likely to be gross inequity of access to facilities unless government 'levels the playing field' – and this process would have to go far beyond linking all schools to the Internet. It would necessarily involve making core software and data sets available at costs feasible even for the most impecunious schools.

The second, related problem is that information providers must still be able to recover their costs and generate profits: in so far as commercial enterprises are the providers or the mediators of information content, it is self-evident that they will seek – and need to be successful otherwise investment will not continue – to generate profits from their activities. The control mechanism for this is typically through the concept of the right of the 'creator' to protect intellectual property and get fairly (not extortionately) paid for its use. How that will work out in a digital world of the type described is not yet clear – but this is *the* central problem.

It may be argued that the advent of local computing power and good telecommunications, allied to the availability of information garnered by millions of individuals across the world, renders this a non-problem – new forms of information supplier will spring up and information will come to resemble free 'shareware'. But this seems inherently unlikely on any major scale: good-quality, nationally consistent geographical (or other) information is not provided by a set of amateurs but by organisations with suitable resources and skills. This was manifested dramatically in the 1986 Domesday project when, despite huge efforts by the BBC and the other partners plus the passionate involvement of most schools in the country, large areas of Britain were uncovered and (inevitably) the quality of what was obtained was somewhat variable. In total, the Herculean efforts of school teachers, pupils and many others produced data sets of only a few percent of the size and subject breadth of those produced by government bodies. This situation is compounded by the changing views on the nature of government. Quite apart from any commercial interests involved, in Britain (and in a rising number of other countries) government policy is now pressing public-sector information providers to recover more and more costs of the collection and assembly of data. This has potentially serious implications for schools unless differentiated costs are charged to different customers (as practised by the Office for National Statistics, some Natural Environment Research Council (NERC) bodies and even by Ordnance Survey).

The obvious conclusion from all of this is that the problems of access to 'content' may be greater than those of access to computing equipment in schools. The latter will almost certainly be eased through the arrival of still cheaper and more powerful computers or personal digital assistants, ultimately providing a 'one-per-pupil' environment. Hopefully this will spurred on by funding initiatives from government, anxious to keep Britain's future work force up to speed in a global business market. In this regard, geography and the enquiry-based and analytical skills it cultivates, linked with computing and IT, will continue to have a direct relevance to the world of work: the approaches we geographers have been advocating for the past two or three decades are now becoming recognised as business-critical, especially within the services sector. Given this, we believe that it is beholden on all geographers to grasp the potential offered by the new technologies to complement and develop core geographical principles, and to pass these skills on to young people – thereby simultaneously keeping the subject vibrant and relevant and ensuring our students are well prepared for the world they will encounter. Given geography's need for 'real-world' data, this places something of an onerous obligation on public-sector information providers like OS, but they too must make their contribution.

Acknowledgements

Thanks are due to Chris Kington for his helpful comments on a draft of this chapter.

Chapter 3

GIS in school education

You don't necessarily need a microcomputer

David R. Green

A 'GIS' (Geographic Information System) is still probably the most significant and important 'buzzword' presently associated with geography, especially for those of us involved in higher education. GIS currently means 'big things' to business, local government planning offices, and last but not least, higher education. Above all it means money, sales of computer hardware and software, research contracts, training courses, and the likelihood of long-term employment for many geography students graduating from universities and colleges of higher education. The growing need for spatial information in all walks of life means that the geographer has arguably rapidly become the centre of the GIS 'world'.

For many geographers, GIS can be viewed, at least in the foreseeable future, as the next best thing to 'sliced bread'. Geography has entered the world of 'high-tech' microcomputers, digitisers, scanners, databases, statistics, graphics, colour printers and plotters, as it never has done before. In many ways, the rapid development of GIS has been responsible for breathing some new life into the discipline of geography, and has undoubtedly brought it once again to the forefront of both public and commercial attention. It has also provided one way by which the business community can forge new links with academia and vice versa; an important step forward in the new structure and role of higher education.

Not surprisingly, and hopefully this is a reflection of the significance of GIS to geography as a whole, the associated terminology is slowly filtering down the education hierarchy. The National Curriculum Interim Report of the Geography Working Group specified as one of the attainment targets for school-level geography students, the need for geography students to be able:

> To synthesise information from various map sources, including atlases, to establish associations between observed patterns (e.g. bring together information from general Ordnance Survey maps, land-use maps, soil maps, and conservation maps) construct a composite map for specified purpose; by overlaying separate distributions of thematic data and evaluate its effectiveness as a geographical information system (e.g. use

transparent overlays or appropriate IT (information technology) to combine data and show inter-relationships).

(National Curriculum Working Group, p. 67)

Clearly GIS has been regarded as an important part of future geography education, both on its own and in association with other components of the National Curriculum such as IT and environmental studies.

But, beyond a brief mention in this short paragraph of the National Curriculum document, what really is GIS at the school level? What necessitates any interest in this subject at the school level? What would serve to differentiate it from GIS taught at higher levels in the education hierarchy? And what would GIS education in schools entail? At the present time there is really no detailed set of answers available to any one of these questions. GIS is not yet a part of the school geography curriculum in its own right, and its significance at this level in the education hierarchy has not really been fully examined or assessed to date. Exactly what GIS would entail with regard to resourcing, training and teaching has also only been examined to a limited degree. In fact this is still somewhat of a puzzle at higher levels!

Brief conversations with school teachers reveals somewhat limited enthusiasm for the teaching of GIS as part of the geography curriculum. Many of these reactions seem to arise from the general misconception that education in GIS requires expensive computer systems and software programs that are neither available nor affordable on the limited budget of most schools. In addition, some teachers perceive a requirement for extensive retraining often in areas they feel are slightly alien to their geographical background. This apparent lack of interest has also been compounded by the fact that there has been, and still is, very little in the way of suitable literature available to promote GIS awareness and knowledge for the school teacher. There have, for example, been relatively few papers in educational journals which have specifically focused on GIS in geography at the school level.

A large part of the problem facing teachers and educators in deciding whether GIS should be taught or not in schools therefore rests upon:

- developing an awareness of GIS
- providing a definition of GIS
- showing how it relates to the curriculum geographers teach now in schools
- defining the level at which GIS should be taught in schools in relation higher education, and how GIS can be taught practically both *with* and *without* information technology

The objectives of this chapter, therefore, are firstly, to provide an outline of GIS to aid the school teacher, secondly, to indicate where it might slot into the geography curriculum and why it might be important, and finally, to examine some of the ways in which GIS can be taught both with and without information technology at different levels of school education. A number of outline examples are used illustrate.

In all probability it is highly likely that GIS or components of GIS are already being taught in schools, though they may not necessarily be identified as such under the heading GIS. If this is the case, then the paper will serve to raise awareness, establish connections between geography and GIS, and stimulate some responses from the teaching community.

What is GIS? Problems of definition for schools

What is GIS? Three definitions are provided below:

- An integrated spatial database management system using database technology and spatial databases
- A computer-based system comprising one or more databases together with a set of tools for collecting (capturing) storing, checking, editing, retrieving, integrating, manipulating, transforming, analysing and displaying geographical referenced data
- A powerful set of tools for collecting, storing, retrieving at will, transforming and displaying spatial data from the real world for a particular set of purposes

Examining these definitions, and many others provided in the GIS literature, it is very hard to find one at present which is really suitable for use at an introductory level required for the geography teacher or for school pupils. Furthermore, most of the current definitions are geared to GIS in the context of IT, making reference to microcomputers, hardware, software, databases, database management systems, and digitising. These definitions, whilst perhaps suitable at the upper end of school education, are hardly appropriate at the introductory level or for schools where IT is currently unavailable. They are also perhaps a little too technical. One solution to this would be to create a new set of GIS definitions which relate to the different levels of attainment of the school pupil e.g. at different stages in their learning, vocabulary, comprehension and grasp of concepts. They would need to be written in such a way as to avoid becoming confusing at a later stage when other more advanced definitions, including elements of information technology, were subsequently introduced. In the meantime, a more detailed and descriptive version of figure 3.2 would probably be more useful at the introductory level.

Undoubtedly, the term GIS is currently associated with computer hardware and software. Together these provide the means for data input, storage, manipulation, processing and output. Computer technology has given the geographer an efficient and very powerful 'desk-top' tool for collecting, storing, manipulating, analysing and outputting both spatial and temporal geographic data/information at a variety of different scales. In addition, the technology has provided a capability to acquire knowledge, understand and plan our environment in ways never before possible. The power and processor speed of the microcomputer allows the manipulation of large data sets which would not be practical using, for example, a calculator or a pencil and paper. The display capabilities of a microcomputer also permit rapid visualisation of data in the form of different types of maps or graphics, while the data manipulation capabilities permit almost instantaneous calculations, statistical analyses, combinations and re-combinations of variables, and repetitive mapping of variables. The storage capacity of a microcomputer also facilitates the archiving of large amounts of data in databases, data which can be rapidly accessed or queried, integrated or used in conjunction with other data.

In reality, many of the fundamental concepts of computer-based GIS are, in fact, not all that new to a traditionally trained geographer, and at the most basic level, the major difference lies only with the working environment: that is, computers and associated hardware and software technology versus pencil and paper and/or plastic overlay transparencies. The technological difference, however, is more significant in a number of ways, not least in terms of the quantity of the data a computer-based GIS can handle, but also in the range of data manipulation and processing capabilities available, some of which are unique when compared to what can be achieved using manual techniques. In addition, it is possible to combine and manipulate far more variables or themes with the aid of the computer than is possible with transparent overlays.

In manual GIS, the brain is the 'controlling system' which identifies, gathers and collates the information in many forms. The disadvantages of such a system are that the approach is relatively slow, unpredictable, incomplete and inventive. With a computer-based GIS, the program structure is used for data acquisition, analysis and display. The approach is digital, fast, standardised, and complete but uninventive. The brain, however, is still required for task analysis and data selection, and in this sense this is where a manual approach can be most useful. A clear indication of the advantages and disadvantages of the 'old' versus the 'new' must therefore be made. These are important for definitions of GIS at the school level. Indeed some of the limitations of manual techniques make them ideally suited to school-level education.

As Unwin (1989) observes, GIS has very little to do with the acquisition and operation of 'flashy' computer systems:

> While the new technology indeed offers geographers a very powerful and useful new tool, the basics of GIS are still the geographic fundamentals such as spatial co-ordinates and referencing systems, map projections, thematic and topographic mapping and survey, spatial operations such as sieve mapping, contour mapping and spatial interpolation.
>
> (Unwin, 1989: 2)

The message of importance to be conveyed here is that in order to be able to use a computer-based GIS correctly, a sound understanding of geographical principles must first be developed. At an introductory level, therefore, the collection, analysis and communication of geographical data does not require immediate access to the processing, storage and data-manipulation capabilities of a microcomputer system. Indeed, even where IT is available, this does not necessitate its use as a replacement for developing a sound understanding of geographical principles.

Why might GIS be an important component in school geography?

The ever-growing demand for data and information about virtually all aspects of the environment, the adoption of information technology, and its impact on society mean that few people will be spared the need to make use of spatial information in the future. IT is part of our society and one that is growing in use very rapidly. Geographers are unavoidably part of this IT revolution. Geography is about the Earth, the human and physical environment; it is about different places at different scales; it is about patterns and relationships; it is about space and time. The new technology and the future significance of spatial data cannot therefore be ignored in the future education of a geography student.

If the basics of GIS are already being taught at school under a different heading why is there a need to emphasise GIS? What does it offer the geography student that is not already offered at present? Some possible answers to these questions are:

- GIS can provide a valuable focus for geography. It allows a student to piece together apparently unrelated data into a coherent whole. With GIS, pupils are able to see how geographic tools, techniques and skills can be used to collect, manipulate, integrate, analyse, and display geographical data/information. School pupils need to be able to understand why, how and when to use certain techniques and concepts, and to understand what happens to geographical data when they are manipulated and analysed. GIS also has the potential to bring geography theory to life through practical applications e.g. using co-ordinates, spatially referenced data, and maps. It is important for school pupils to be able to use geographical skills practically rather than just to be told about them.
- GIS allows students to make full use of knowledge and techniques from other National Curriculum studies e.g. mathematics and information technology (or computing studies). Where computer facilities are available, simple GIS software can utilise the IT/computing studies skills acquired in school for use in the context of geographical applications. GIS is also important in relation to environmental education, a key area

in the future of geography education. Furthermore, it could even help to develop skills in logical thought, question-and-answer processes, and problem solving.

- By introducing GIS at an early stage it is possible to develop a progression in learning about the use of spatial data in geography right the way through to higher education. This benefits both the education of the student as well as educators in higher education courses.

- GIS techniques can provide a student with a 'hands-on' insight into how geographical data can be used as the basis for planning and decision making.

- As education moves towards more enquiry-based learning strategies, students' knowledge of databases must be developed such that they take a questioning attitude into the workplace in support of specific and transferable skills (Coggins, 1990). GIS places databases in a geographical context.

GIS in the education hierarchy

In order to establish what might be involved in teaching GIS at the school level it is worth briefly examining GIS throughout the education system.

Postgraduate

In recent years, the availability of diploma and MA/MSc degree courses in GIS has grown quite rapidly at the postgraduate level. Courses are now offered, for example, at University of Edinburgh, Birkbeck College, University of Leicester, the University of Salford, University of Nottingham, the University of Leeds, University of Luton and Kingston University. These aim to teach students the theory of GIS and, in addition, to provide vital hands-on experience with proprietary GIS computer software and hardware as it would be used in the 'real world'. To some extent this also involves a measure of training in the use of certain equipment, although education, and not training as such, is the key objective of these degree and diploma courses. More recently, the traditional GIS courses have been joined by distance learning e.g. UNIGIS and Internet-based courses (Birkbeck College).

The adoption of GIS education at this level has, however, been relatively slow and there were a number of educational difficulties cited by Unwin (1989) which may have accounted for the relatively slow progress in introducing GIS courses at this level. These are problems of defining training versus education in GIS, syllabus or curriculum definition, technological dependence, and differing student backgrounds. A suggested syllabus was provided by Unwin (1989) (see figure 3.1). Attempts to overcome such problems have been made by the NCGIA in North America through the rapid assembly of a curriculum designed to 'speed up' the learning process.

Prior knowledge
basic cartography
computing
statistical analysis
data handling
data management

Context of GIS
definitions and component subjects
commodification of information
GIS capabilities

Cartography/spatial analysis
typology of spatial data
georeferencing
map projections/transformations
2-D and 3-D co-ordinate transformations
spatial concepts
quality of spatial information
point analysis
line analysis
area analysis
surface analysis

Computer realisation
low and high level data structures
raster data models
vector data models
towards expert systems

Operational considerations
hardware overview
data storage
standards
display and output models
proprietary systems

Application of GIS
real world users
international initiatives
decision making
project management
costs and benefits

Institutional issues
data ownership, copyright, legal issues
management and training

Figure 3.1 Unwin's proposed curriculum (Source: provided by Unwin (1989), taken from
the journal *Computers in Geography*, who hold the copyright.)

Undergraduate

At the undergraduate level, option courses in GIS have, to date, been less well
developed. However, the term Geographic Information System has inevitably
crept into the vocabulary of many geography students who opt to take courses
in computer applications in geography, remote sensing and cartography, and/

or topographic and mapping science. Increasing awareness of GIS at the undergraduate level is also resulting from a number of other factors: for example, recruitment of younger staff in universities and colleges with a background in GIS and computer applications and/or IT, changes in the education system e.g. movement to modular degree schemes aiming to provide the students with a wider choice of courses, and increasing demand for the use of information technology in geographical applications. Closer links between universities and business have also served to promote awareness of GIS at this level of study through class visits to companies (e.g. at University of Aberdeen visits are arranged to BP in Aberdeen, Masons Surveys in Dunfermline, and Bartholomews [Harper Collins] in Edinburgh) and through videos e.g. videos made available by the Economic and Social Research Council (ESRC) and companies e.g. Environmental Systems Research Institute (ESRI) and Siemens. The increasing availability of computer software and hardware in universities and colleges in the form of GIS tutors (e.g. GIST) and GeoCube have also helped to increase awareness quite significantly. In recent years, the Internet has also provided a great deal more access to an excellent resource of illustrative material.

School

What needs to be taught at the school level and when? Coggins (1990) drew up a GIS curriculum for schools based on Unwin's higher education curriculum. Figure 3.2 outlines the curriculum content. Comparison with figure 3.1 shows that it follows the broad categories set out by Unwin which are: prior knowledge, context of GIS, cartography/spatial analysis, computer realisation, operational considerations, application of GIS and institutional issues. This provides one step towards a general GIS framework allowing for progression from school to university level. The curriculum has been devised in such a way that it encompasses both manual and computer-based GIS at a level which is below that expected of students in higher education. If this curriculum were to be developed in more detail then it would also need to include a progression in GIS education throughout school.

The GIS fundamentals could all be developed from an early age taking into account the expected attainment of the student at that age in geography and other subjects. For example:

- Age 5–7: shapes, size, distance, area, perimeters, direction, slope, models, simple mathematics and practical work
- Age 7–11: distance, direction, reference systems, maps, areas, slopes, steepness, fieldwork
- Age 11–14: characteristics of map types, map projections, aerial photographs, vector to raster, grid references

- Age 14–16: map comparison, distance and direction, contour maps, satellite imagery, projections, databases, geographical enquiry spatial associations, cross sections
- Age 16+: project work

Some potential problems and solutions

Problems

In addition to the problems outlined by Unwin (1989), some of which are also applicable at this level in the education system, the following problems exist which may contribute to an apparent lack of enthusiasm about teaching GIS at school level:

- GIS is often not clearly understood because existing definitions of GIS presented in the literature are too obscure for those not familiar with the associated terminology used in these definitions.
- The subject matter covered by GIS is often unclear. There is no real consensus, as yet, on what is required. Furthermore, although GIS is essentially, as Unwin (1989) states, 'What Geography is all about', the connections linking the two are seldom immediately clear from the literature of higher education.
- In relation to the previous point, GIS has frequently become associated in people's minds with computers. This association automatically triggers a lack of interest on the part of some people for several reasons: not everyone comes to terms with the idea of having to use computers, even if they are not necessary, and the thought of having to undergo additional training particularly to enable the use of information technology often does not appeal, particularly to an older generation. In addition, they may have the misconception that microcomputers are an 'absolute must' before one can teach GIS, or that only very sophisticated and expensive computer software and hardware are necessary.
- In relation to the two previous points, the perception that GIS requires computers and associated peripheral hardware and software has also discouraged the introduction of GIS in schools to some extent.
- There is relatively little at the present time in the way of supporting information about GIS which is suitable for the school teacher and student. While a great deal of information now exists about GIS and is readily accessible to anyone interested, more often than not it is pitched at a completely different level of reader. It usually assumes a prior knowledge of computing hardware, software and GIS, and thereby does not provide a starting point for the intending teacher or student. After all, we all need to start at the beginning when starting a new subject or course!

Prior knowledge
basic map skills
keyboard skills
basic numeracy
data collection, fieldwork
data recording

Context of GIS
geography of areas
data sources
enquiry-based learning

Cartography/spatial analysis
points, lines, areas, surfaces
names, grid references
map projections
maps and images
distance, orientation
scale, accuracy
counts, dispersion
networks, routes
perimeter, area, overlays
relief representation

Computer realisation
keyboard skills
satellite images
enter points and lines
simple software packages

Operational considerations
data entry systems
access to external data
data searches/screen dumps
Domesday/satellite signals

Application of GIS
visits to users
locational games
evaluate geographical questions

Institutional issues
reliability of data

Figure 3.2 Coggins school curriculum (Source: Coggins, 1990)

- There is currently a lack of available data and suitable databases to enable the teaching of computer-based GIS at different levels.
- Much of the computer software now available for GIS has not yet been geared to a school environment, or to the types of computers available in many of the schools. A large number of schools still have BBCs and Archimedes for which there is no commercially available GIS software. While other schools may have RM-Nimbus, and even IBMs or compatibles, and Apple Macintoshes, for which there are now a wide variety of GIS software available, this is not the general trend.

- In terms of cost, the computer hardware and peripherals, e.g. digitisers, printers and storage devices required by some of the GIS packages are usually beyond the school budget. Furthermore, it is unlikely that schools would be able to purchase a large number of complete systems for classroom use and for hands-on experience without assistance.

Solutions

Some possible solutions to these potential problems would be to

- raise teacher awareness through regular articles in geography journals about the use of GIS
- devise a school-level GIS curriculum
- initiate educational seminars, training sessions and meetings at local, regional and national levels
- develop co-operative links with local government, business and higher education for help with teaching materials, databases and software, and perhaps even hardware
- develop GIS teaching and resource packs including databases, overlays, videos, reading materials, lists of contacts, and even a newsletter
- develop software to run on a variety of low-cost microcomputers to act as self-teaching tutors, demonstrations and practical sessions

How GIS might be taught in schools

GIS could be taught at school in a number of ways using both computer-based and manual approaches. This would of course depend upon the age group, the acquisition of appropriate complementary skills and knowledge from other subjects by a certain age, e.g. mathematics, computing studies, the availability of appropriate computer hardware and software, and resourcing. Four different approaches would be possible. These are:

- the use of manual overlay techniques (only)
- the use of microcomputers (only)
- complementary use of manual techniques and microcomputers and/or progression from overlay techniques to computer-based GIS
- the use of a computer-based GIS to produce material suitable for developing transparent overlays

Although a brief section in this chapter will subsequently be devoted to GIS using computer technology, the remainder of the chapter will concentrate on manual overlay techniques as the main approach that would best be used at the school level to teach the fundamentals of GIS.

Transparent Overlay Techniques

Transparent Overlay Techniques are a well established method developed many years ago. According to Steinitz et al. (1976) they were first used in the 1930s, and since then have been widely and successfully used in environmental planning at a professional level. They have also been used by geographers, often in an environmental context (see for example Smith, 1972). In fact, such techniques are still used at present and to great effect. These are what Walsh (1988) terms the 'Manual GIS Approach', and no doubt have already been used, perhaps unknowingly, in the context of GIS by geography teachers for many years.

Simple overlay techniques can provide a very effective teaching tool and can in fact be introduced at an early age and developed over time at varying degrees of sophistication. They have the advantage of facilitating 'hands-on' experience for all students, and give everyone an opportunity to understand what is involved. At the same time they also permit a practical application of knowledge, and reveal graphically the inter-relationships existing between different variables. Overlay techniques also build very effectively upon a student's existing knowledge of grids, co-ordinates, shapes, areas, lines, points, X and Y co-ordinates, as acquired in elementary mathematics. Children learn about regular and irregular shapes e.g. squares, triangles, pentagons, hexagons, points, lines and areas at an early age, and this learning can easily be placed into a geographical context by subsequently allowing students to investigate their immediate surroundings, e.g. the school classroom. For example, they describe the shape and dimensions of the room, the layout, or location and shapes of physical objects such as desks, tables and chairs. It is also possible to develop the concepts of vectors (lines), rasters (grid cells), and polygons (areas), though not necessarily using such terminology. At Stage 4 in the Scottish Primary Education System, school pupils in primary mathematics are taught about graphs with co-ordinates, locations, and maps. This may involve numbering the Y axis and labelling the X axis with letters (A–Z). In addition, they make grids on paper, shade in grid squares (simple rasterisation), learn about rows and columns, locations according to grids, become familiar with joining points together (vectors!), various shapes, north arrows and scale, grid squares representing so many map units e.g. metres and kilometres, areas in centimetres squared, measurement of perimeters, graphs, distances, lengths and heights (topography) from maps, compass directions, angles and scales.

Two examples, both using overlay techniques, are presented in the appendices as tentative ideas developed for two different age groups. Some further 'ideas' are also provided. The aim is to provide a possible indication of what can be done at minimal expense. This approach allows the student to put into practice simple GIS techniques, while still being able to follow the processes by simple hand-checking of any calculations involved starting with the collection through to the analysis stage. A small data set, e.g. a square of $5\,cm \times 5\,cm$ or $10\,cm \times 10\,cm$ divided into 25 or 100 $1\,cm^2$ grid cells, is quite adequate and can be managed both visually and mathematically with relative ease. This type of approach removes the need to make use of microcomputers. A similar approach is often used in introductory university education text-books or software.

There are also a number of ways in which elements of GIS can be developed using simple and relatively cheap approaches with only the requirement for a pencil and paper, graph paper, and some plastic transparencies and/or tracing paper. These exercises can then be developed at different levels suitable for different age groups. Over time, exercises can be provided for a range of different geographical scales, incorporating various geographical techniques, at varying levels of sophistication, and perhaps even including the use of IT, and fieldwork. For example, one might start out by asking students to draw some simple regular and irregular shapes on a sheet of paper e.g. a line, a point, square, rectangle, circle, and a triangle (see for example, figure 3.3). Perimeters and dimensions can easily be measured with the aid of several pieces of string and rulers. The next step might be to transfer these shapes to a sheet of graph paper, adding X and Y co-ordinate axes. Once again some measurements can be made recording the X,Y locations and co-ordinates of the shapes (e.g. a single X,Y for a point, a pair of X,Y co-ordinates for a line, and a string or series of X,Y co-ordinates for a shape/area/polygon). Areas of both regular and irregular polygons can easily be calculated knowing length of each side, and by counting the number of cells of known size. Students can also examine the effect of using different sized grid cells on the accuracy of area estimates and can experiment with the transformation of vector (line) data to raster format (figure 3.4). Here it would be necessary to indicate that large grid cells (rasters) are used only for the purpose of explanation and in reality should be made so small as to be 'invisible'.

Next, one might encourage students to draw a simple plan map of their classroom, or alternatively, at a later stage, of their local surroundings either at home or school, mapping the locations of familiar objects and features onto paper e.g. areas comprising different surfaces (tarmac, concrete, soil, grass) areas of grass, flowerbeds (see for example, figure 3.5 (taken 'out of context'

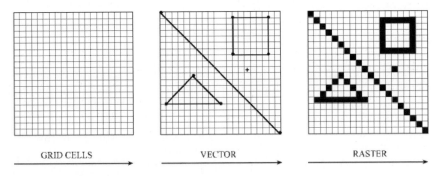

GRID CELLS   VECTOR   RASTER

Figure 3.3 Simple regular and irregular shapes on a sheet of paper

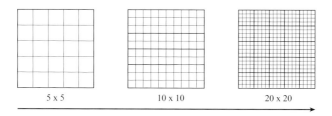

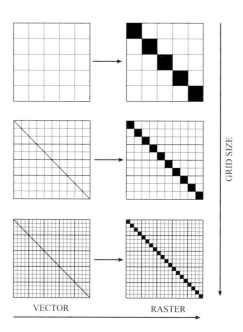

Figure 3.4 The effect of different sized grid cells on the accuracy of area estimates, and the transformation of vector (line) data to raster format

from Maguire, 1989)). In a similar fashion to that described above, students can then grid the information to develop raster overlays (or layers of information) separating out, for example, the different surface types. In both cases, children can begin to learn about databases. This is achieved simply with the aid of small sets of file cards considered to be records containing fields with data about an object e.g. its measurements, colour, and shape. As groups, pupils may then use these in other class exercises or projects (see appendix 1). In the above example, pupils are able to produce a simple map database (the raster grid cells) with the surface type being the surface attribute for each map unit.

At a more advanced level, the geographical environment selected might comprise a larger area, e.g. a village, where children would be assigned a small

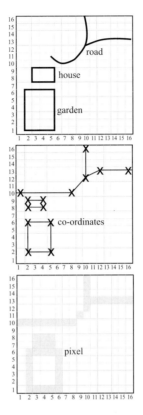

Figure 3.5 A simple plan map of local surroundings either at home or school (Source: redrawn from Maguire, 1989)

area on a map, and then be required to acquire data and information about that area e.g. rivers, transport routes, village morphology etc. The map used, however, need not necessarily be of a real area and can just as well be fictional. Maps of different scales provide different detail. Selected data and information can be read from a map and drawn onto different transparencies e.g. rivers on one overlay, roads on another. Certain questions (queries) can then be asked of the pupil based upon the information they have selected e.g. at what locations do the 'A' roads cross rivers? What is the distance between X and Y? How far by road from farm A is the nearest town? How large is the area covered? What percentage of the area is forested? This information might also be mapped onto a separate layer, or placed in a simple database. Databases can be created using different levels of attribute detail using filecards. Changes detected between observations made at one time and those made at another time can also be made, e.g. simple change detection and updating. This might be aided through the use of aerial photography of an area taken on different dates.

At a later stage, overlay techniques might be developed at a more sophisticated level perhaps by placing transparent overlays on aerial photographs, and even satellite images, to show how different types of data can be acquired from different sources and combined with other data derived from maps (figure 3.6). Some knowledge of the detail of data available from different scale maps and images can also be developed at this stage together with mention of the problems of distortions inherent in aerial photographs. Similar exercises can incorporate the use of perspective views, contour maps, and block diagrams for a variety of different geographical studies of different places, at different scales e.g. local, regional, national and even global. Problems of using one source of data with another can also be discussed and demonstrated. Small projects can be developed e.g. a map representing

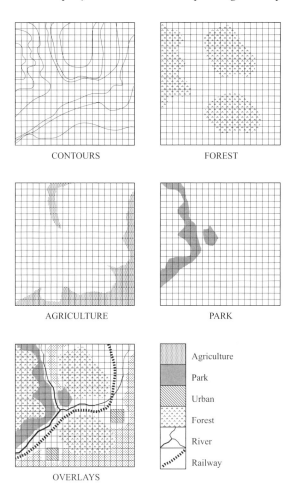

Figure 3.6 Separate themes combined into one using overlay techniques

harvestable timber in a forest overlain on a soil map, and sites identified for which special management efforts or restraints are required, e.g. sites with low volumes of timber and unstable soils might be avoided for commercial harvest (Daniel et al., 1977). Another project topic might investigate the planning of a bypass around a town using overlays containing data and information about geology, soils, agricultural land, existing transport networks, rivers, towns and sites of historical or ecological interest. This might involve mapping different variables on separate overlays, rasterising (gridding), and the provision of numerical weightings for each variable and grid cell (see, for example, Walsh, 1988). A multiple overlay map with combined weighting e.g. value of variable 1 in grid cell (1,2) plus value of variable 2 in grid cell (1,2) ... plus value of variable n in grid cell (1,2) would be created by overlaying each individual map, adding up the weightings and plotting the respective boundaries.

With the introduction of IT, if available, such exercises as those described above could also be placed on a simple computer-based GIS by a teacher to facilitate comparisons with the manual approach. Advantages and disadvantages of both approaches could then be examined and assessed, introducing the student to some of the techniques for maximising the use of shading and colouring to allow the mapping of information on overlays more effectively e.g. combining colours and shading patterns, and to show the disadvantages of overlay techniques when there are a large number of overlays. Alternatively, a teacher familiar with a computer-based GIS could produce overlays on the computer system for use in manual exercises.

Environmental quality assessment

The next example, also using overlay techniques, is probably more suitable for an older age group, perhaps as an illustration for a potential project. It could also be implemented on a microcomputer-based GIS for comparison if needed.

This exercise builds on the allocation of weightings for environmental quality assessment and is taken from the work of Smith (1972). A 1 km Ordnance Survey grid forms the sampling base for the Black Isle study area (Ross and Cromarty, Scotland). This approach is both objective and also convenient for data storage. A more detailed study could be made by dividing each 1 km grid square into four.

Six environmental qualities are investigated. These are habitat diversity, land capability, scientific and ecological interest, topography, landform, and historic buildings. Habitat diversity consists of twelve variables involving ground cover e.g. arable ground, rotation grass, coniferous woodland, deciduous woodland, rough grazing with scrub, rough grazing, moorland, freshwater stream or lochan, tidal water, loch, shingle beach, sand beach, slat marsh and sandflat. The Macaulay Land Capability Classification, defining the food-producing value of agricultural land, is noted for each square having

>50% land area in either Class I and II or III. Topographic values can be acquired from a 1 : 2,500 scale OS map with contour intervals of 25 feet. Micro-landforms can be assessed from fieldwork. Ecological Interest is based on the occurrence of a site of scientific interest in a grid square. The Statutory List issued by the Secretary of State for Scotland (1971) for Ross and Cromarty provides information on the presence of listed buildings.

For each of the six environmental quality variables an overlay of grid squares is made for the study area and in each square a number of points is allocated according to the following scheme:

Habitat diversity	rating from 1 to 12 points (sum of occurrence of each category of ground cover as above)
Land capability	over 50% Class I and II = 10 points
	over 50% Class III = 6 points
Ecology	4 points for presence of site of scientific interest
Topography	> 500 feet relief amplitude = 3 points
	250–500 feet relief amplitude = 2 points
Landform	1 point for each additional micro-landform
Building	1 point for listed building.

The individual maps or overlays are then combined and a total 'points map' or weighting composite is created. This is a map of 'environmental quality', a new variable which has been derived by combining a number of other variables using measurements acquired in fieldwork or from map data sources.

Limitations of the overlay technique

While overlay techniques are potentially very useful in teaching some of the principles and concepts of GIS, it must also be remembered that they have a number of distinct disadvantages. These are important to recognise and to convey in GIS education. Such disadvantages, for example, are limitations in the number of overlays that can be combined. Mechanically it becomes difficult to combine a large number of transparent overlays in terms of visual graphics. The approach is also often more qualitative than quantitative. In many ways the approach is 'map specific'. Different combinations of map overlays are time-consuming to create, often necessitating that the entire map overlay be redrawn.

Microcomputers and GIS: complementary and supplementary but not necessarily a replacement!

There has so far been very little in the way of software specifically aimed at teaching GIS to students of a school age. In some ways this is not surprising given the relative 'newness' of GIS in school education. However, there are now a number of software packages available, e.g. AEGIS, IDRISI, and PMAP as well as MapInfo™, ArcView and ArcExplorer initially aimed at the higher education market which can be (and are being) used quite satisfactorily at the upper school level, provided students have acquired a firm grasp of the basics of GIS beforehand. Use of these packages requires a teacher to be familiar with the operation of hardware and software.

A number of years ago access to the Acorn Domesday system and any microcomputers with spreadsheet and database software also offered the teacher a practical and useful alternative to specialised GIS packages. These could be used instead of, or in addition to, specialised GIS packages and at the same time would also serve to provide an intermediate 'step-up' from the manual GIS techniques described earlier e.g. when teaching databases and the fundamentals of database enquiry.

Unfortunately, although spreadsheets are an attractive and useful alternative, especially when funds for software are limited, they are not really a long-term solution worth considering in place of the newer and cheaper GIS software now becoming available. However, on the positive side the value of using a spreadsheet can also be viewed in terms of its potential uses in other areas of the curriculum e.g. simple model simulations, likely pupil and teacher familiarity, and availability.

Maguire (1989) and others (see Maguire, 1989) have reported on the value of the Domesday system for undergraduate education in GIS but this type of system, through not strictly a true GIS could and has been equally useful when applied at the school level. Indeed it is schoolchildren who were involved in collecting some of the data for the Domesday databases. Access to this type of system did provide a useful intermediary link between geographical knowledge with IT and database enquiry. However, it is no longer relevant as it is long since out-of-date.

Campbell (1989, 1990) has provided several useful examples of how a standard spreadsheet can be used to input, store, query and map geographical data. A modified example of his ideas is provided as an example in the appendix (see also figures 3.7 to 3.10)

In a short paper reviewing the IDRISI GIS software package, which retails for approximately $75–200 complete with manual (see http://www.clarklabs.org), Maguire (1990) notes that it can be used to teach the principles of GIS to students of all ages from school to MSc level, and perhaps more importantly, to teach substantive areas of geography, for example, biogeography and historical geography. This is very encouraging and indicates clearly that with the current changes in the geography curriculum, the requirements of IT at school levels coupled with the availability of cheaper software and hardware, and links between education and business, GIS can easily be incorporated into the geography curriculum at an early level in the not too distant future.

Microsoft Works

File Edit View Insert Format Tools Window Help

Arial   10   B / U   Σ

C3   "Abb

GIS_SP"1.WKS

	A	B	C	D	E	F	G	H	I
1									
2		Place	Abbrev	X-Axis	Y-Axis	Hsn-Ass%	Loc Ath%	Own Occ	PrR+Oth%
3		Abbey W	Abb	466	785	1	34	58	8
4		Arsenal	Ars	438	788	1	87	9	4
5		Burrage	Bur	440	782	9	47	34	10
6		Eynsham	Eyn	465	794	0.1	72	26	3.3
7		Glyndon	Gly	443	788	5	63	23	9
8		Herbert	Her	434	775	8	44	39	9
9		Lakedale	Lak	451	784	5	22	60	13
10		Nightingl	Nig	435	780	0.1	83	0.1	17.4
11		Plumstea	Plu	454.2	772.4	14	14	60	12
12		St. Marys	StM	431	789	2	93	1	3.2
13		St. Nicho	StN	463	779	5	20	63	13
14		Shrewsbr	Shr	439	770	10	7	75	8
15		Slade	Sla	462	775	2	29	62	8
16		Thamesm	Tha	465	803	0.1	94	5	1.1
17		Woolwic	Woo	430	782	4	55	20	21
18									
19									

Figure 3.7 Input data (Source: after Campbell, 1989)

Microsoft Works - [GIS.W

File Edit View Insert Format Tools Window

Arial   1U

K16

	A	B	C	D	E
18					
19		Abbrev	X-axis	Y-axis	
20		Ars	438	788	
21		Gly	443	788	
22		Nig	435	780	
23		StM	431	789	
24		Tha	465	803	
25		Woo	430	782	
26					
27		Bur	440	782	
28		Eyn	465	794	
29		Her	434	775	
30					
31		Abb	466	785	
32		Lak	451	784	
33		Plu	448	780	
34		StN	463	779	
35		Shr	439	770	
36		Sla	462	775	
37					

Figure 3.8 Extracted data part 1: map co-ordinates (Source: after Campbell, 1989)

	E	F	G	H	I	J	K	L
18								
19		Abbrev	X-axis	Y-axis	Y-axis	Y-axis		
20		Ars	438	788				
21		Gly	443	788				
22		Nig	435	780				
23		StM	431	789				
24		Tha	465	803				
25		Woo	430	782				
26		Bur	440		782			
27		Eyn	465		794			
28		Her	434		775			
29		Abb	466			785		
30		Lak	451			784		
31		Plu	448			780		
32		StN	463			779		
33		Shr	439			770		
34		Sla	462			775		
35								

Figure 3.9 Extracted data part 2: map co-ordinates separated into columns (Source: after Campbell, 1989)

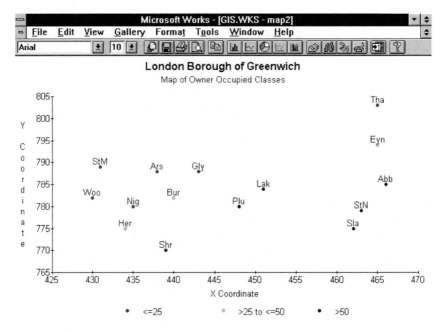

Figure 3.10 The 'Map' (Source: after Campbell, 1989)

A consideration that would need to be taken into account is the ease of use of the software by teachers. It is worth noting, however, that software and hardware are becoming increasingly user-friendly, and with good software tutorials accompanying many pieces of software, excellent manuals e.g. IDRISI and software reviews, together with the fact that children readily adapt to a computer environment, GIS in a computer environment should not really be beyond the grasp of any geography teacher. Additional factors might be provision for some training, the establishment of simple GIS exercises, creation and availability of suitable databases, projects, and data sources.

To reiterate an important point though, no matter how one chooses to use the IT there is still a need to develop a progression in manual GIS throughout school education to provide for a sound knowledge of the principles of GIS as a vital prerequisite for computer-based GIS. This would avoid students merely 'button pushing', which unfortunately it is all too easy to do, leading to the acquisition of an end product, e.g. a map, without knowing whether it is correct or what it means.

To some extent this sort of problem also arises from using pre-printed tutorials, which whilst useful are often all too easy to complete without a student having to think about what they are doing. This is why it is important to introduce the theory and simple 'hands-on' approaches with pen and paper, and even spreadsheet software, to ensure that pupils get to grips with the fundamentals before moving to menu-driven software which is easy to use, but often does not require very much thought on the part of the student.

Recent developments

Whilst it is important at the outset in GIS education to encourage development of a theoretical background through a combination of the provision of the fundamentals together with application of the knowledge to develop understanding, especially prior to starting on computers, information technology can be introduced by a teacher as a means to illustrate the practical use of GIS software, and to show examples. In many ways the approach taken by the OSNI CD-ROM is a good example, whereby computers can be used to run the CD-ROM provided, which itself provides plenty of insight into the theory and applications of GIS, but a step-by-step player version of applications using ArcView also helps to reveal the ways in which GIS can be used. Similar examples can be revealed through the use of demonstration diskettes which offer slide shows of data and applications which can simply be examined. Worksheets to accompany these demonstrations can also help the pupil.

Conclusions

While GIS, by definition, clearly involves the use of computer information technology, introduction at an early stage of the principles and techniques of geographic data handling necessitates that one use manual approaches, irrespective of the availability of appropriate hardware and software. The use of overlay techniques, as described above, can be used quite satisfactorily at a number of different age levels to develop and teach the concepts and principles of GIS both at a theoretical and practical level and quite inexpensively, allowing all pupils to have hands-on experience of the use of computer information technology. Introduction at an early stage of the principles and techniques of geographic data handling necessitates that one use manual approaches irrespective of the availability of appropriate hardware and software. The requirement for computers can, therefore, be left until later, if computer systems are available, or avoided altogether since they are not absolutely necessary.

If IT is to be incorporated in school GIS education then it could be developed progressively through demonstrations, teacher-aided use of simple software packages dealing with certain aspects of GIS, e.g. spreadsheets, databases and GIS software, and finally hands-on experience with simple GIS software, e.g. ArcView. It would be possible for a teacher to implement, for example, a series of exercises using manual overlays on a computer system using relatively inexpensive software referred to earlier. This could then be used to demonstrate to pupils the advantages and disadvantages of computer systems, e.g. the pitfalls of too much button-pushing without understanding the principles. The lack of data sets or perhaps even lack of appropriate computer peripherals such as digitisers and video digitisers could even be overcome by closer ties between schools and higher education where such hardware is readily available.

There is, I believe, considerable potential and merit attached to the introduction of GIS within geography education. The benefits are many if a curriculum were to be developed with sufficient care and thought. Promoting the use of manual techniques at this stage in geography education is not to dismiss the value of information technology in a geography environment, but rather to place an emphasis on the importance of developing a thorough understanding of what is involved in the collection, analysis, and use of geographical data and in such a way that it is beneficial to the student's overall future education. GIS could be introduced not so much as the core of geography but as a new application. While much of GIS is what geography teachers teach now, it could nevertheless provide a very valuable stimulus for geography as a subject area, providing a focus for the subject matter, which would help to explain in a theoretical, graphical and practical way many geographical concepts. In many ways it would also help to provide cross-disciplinary links and make better use of other subject matter e.g. mathematics, environmental education and information technology in a

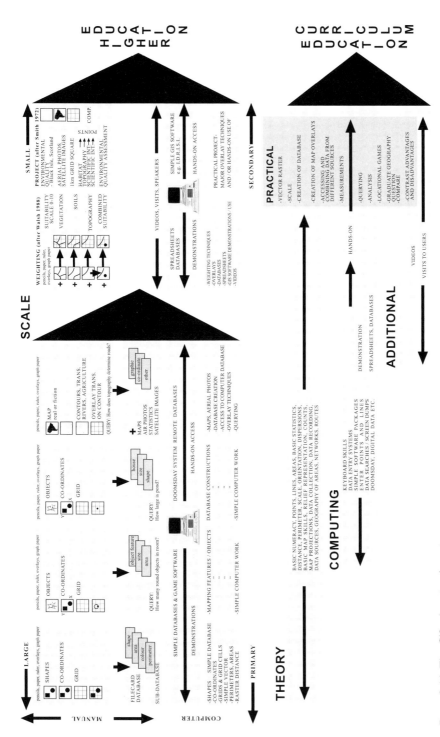

Figure 3.11 The GIS educational continuum

geographical context, providing an important new and worthwhile foundation for geography students and, if developed throughout the school curriculum, a degree of continuity from primary through to higher education and beyond (see figure 3.11).

One should not underestimate the ability of children to deal with the basic concepts of GIS and the practicalities, both in relation to manual and computer techniques, nor should geography teachers be fearful of teaching GIS. Indeed, by not offering GIS as part of the geography curriculum, one could deprive students of many opportunities, not least an up-to-date education in geography. So, maybe GIS in its simplest form is not really all that new, all that incomprehensible, or indeed all that difficult to teach at the school level.

Acknowledgements

Dr J.S. Smith, University of Aberdeen, for permission to use his Environmental Quality Assessment work from 1972.

David J. Maguire for a figure borrowed from his book *Computers in Geography*.

Michael Wood, University of Aberdeen, for reading over this package and providing some very thought-provoking comments, some of which I have managed to incorporate into the overall structure.

References

Campbell, N. (1989). 'The Spreadsheet as a Low Cost Geographical Information System', Paper presented at a Workshop on the New Technology and Spatial Information Systems, University of Nottingham, January, 16p.

Campbell, N. (1990). 'The Spreadsheet and the Political Economy of GIS', Paper presented at University of Leicester.

Coggins, P.C. (1990). 'Horses for Courses: Education in GIS', Mapping Awareness '90 Conference, Blenheim, London, pp. 19-1–19-5.

Daniel, T.C., Anderson, L.M., Schroeder H.W., and Wheeler, L. (1977). 'Mapping The Scenic Beauty of Forest Landscapes', *Leisure Sciences*, Vol. 1(1), pp. 35–52.

Maguire, D.J. (1989). 'The Domesday Interactive Videodisc System in Geography Teaching', *Journal of Geography in Higher Education*, Vol. 13(1), pp. 55–68.

Maguire, D.J. (1990). 'Software Review: IDRISI', *CTICG Newsletter*, No. 1, Spring, p. 4.

National Curriculum Working Group. *Interim Report*, Department of Education and Science, The Welsh Office, 99p.

Scottish Primary Mathematics Group. (1984). *Primary Mathematics: A development through activity. Stage 4. Textbook*, Heinemann Educational Books, London/ Edinburgh, 115p.

Smith, J.S. (1974). 'Development and Rural Conservation in Easter Ross', *Scottish Geographical Magazine*, Vol. 90(1), pp. 42–56.

Steinitz, C., Parker, P. and Jordan, L. (1976). 'Hand-Drawn Overlays: Their History and Prospective Uses', *Landscape Architecture*, September, pp. 444–5.

Unwin, D.J. (1989). 'Geographical Information Systems and School Geography: Some

Essentially Random, but Cautionary, Thoughts', Short Paper presented to the GA Working Party.

Unwin, D.J. (1989). 'A Syllabus for Teaching Geographical Information Systems', *Research Report No. 2*, Midlands Regional Research Laboratory, 13p.

Walsh, S.J. (1988). 'Geographic Information Systems: An Instructional Tool for Earth Science Educators', *Journal of Geography*, January–February, pp. 17–25.

Further reading

Fisher, P.F. (1989). 'Geographical Information Systems Software for Teaching', *Journal of Geography in Higher Education*, Vol. 13(1), pp. 69–80.

Geographical Association (1989). *Some Issues for Geography Within the National Curriculum*, Papers submitted by the Geographical Association to the National Curriculum Geography Working Group, (Ed. S. Catling), September, 90p.

Green D.R. and McEwen, L.J. (1990). 'GIS as a Component of Information Technology Options in Higher Education Geography Courses', in *AGI Yearbook*. (Eds M.J. Foster and P.J. Shand) pp. 287–94, Taylor & Francis, London.

Green, D.R. and McEwen, L.J. (1989). 'GIS as a Component of Information Technology Courses in Higher Education: Meeting the Requirements of Employers', in Proceedings of the AGI National Conference, Birmingham, June 11–12.

Haines-Young, R.H., and DoornKamp, J.C. (1989). Eds, 'Geographical Information Systems for Local Government', *Seminar Lecture Notes*, No. 4, 100p.

McHarg, I.L. (1969). *Design with Nature*, Natural History Press, New York.

APPENDICES

The contents of these appendices have been put together to stimulate ideas rather than being examples of finished exercises.

Appendix 3.1

Task	Where are People in the Classroom?
Tools	Some filecards (plain or coloured)
Method	Once you have drawn a map of your classroom showing the location of everyone's desk, create a small database using a set of file cards e.g. just like a card file system in a library
Definition	A database is a collection of information relating to a particular subject

e.g. using the alphabet (A–Z) mark a card with each letter of the alphabet indicating first letter of LAST NAME of a person. LAST NAME is the KEY field. Then collect some details for each person in the class e.g. first name, height, colour of hair, desk number in class. Using one card per person mark this information on the card:

LAST NAME	FIRST NAME	INITIAL	HEIGHT	COLOUR HAIR	DESK NUMBER
SMITH	FRED	Y	5'6"	BLACK	D3

etc.

File the card under the appropriate LETTER of the Alphabet e.g. S in this case. Each card is known as a RECORD. Each fact about the person is known as a FIELD. How many records do you have? How many fields are on each record?

You might then create a second database but this time using the DESK NUMBER as the KEY field and include only two fields on each card:

DESK NUMBER	LAST NAME	FIRST NAME	INITIAL
D3	SMITH	FRED	Y

Make a card for each desk number.

Try querying the database. Who is sitting at Desk Number D3? Answer someone with the last name SMITH whose first name is FRED and whose initial is Y. You might then want to know more about the person. So using the first database check out the name SMITH under S with additional identifying factors of FIRST NAME ... FRED, and INITIAL ... Y. What colour hair does FRED have? What height is FRED SMITH? and so on.

The types of database constructed could be more complex and involve more geographical data. For example, using a map of postcodes as a reference you could create a small card database for the class, a record for each pupil, and subsequently perhaps plot the home location of each person on the map, thereby showing how postcodes work.

Pupils could add and delete records from such a database, even update records, and create reports. The part of a Database Manager (DBM) could also be worked in.

Single-file databases, multi-file databases, spatial databases, text databases and graphic or picture databases could all be built around this type of filecard structure, and developed at different levels and detail. A gazetteer could be used with an atlas – using name of place, leading to latitude and longitude and eventually location of the place on a map. Databases could also be created beforehand on sticky-backed paper and used in a similar way.

Appendix 3.2

Task　　　　Mapping Your Favourite Room!
Tools　　　Pen, coloured pencils, paper (tracing and graph), ruler, tape measure, your feet!
Method
1　Select your favourite room at home.
2　Measure the length and the width of the room using either a tape measure or the length of your foot.

3   Draw a rectangle on a sheet of graph paper. Make the longest side equal to the longest length of the room you have measured, and the shortest side equal to the shortest length (to the nearest foot). Use 1 grid cell to represent 1 foot.

4   Select 5 objects from the following list chair, table, bed, lampshade, dresser, cupboard, sideboard, sofa, and measure the size of these objects (length and width). Using the lower left corner of your rectangle as your origin measure the distance in the X and Y direction of each object from the origin. Record these measurements. Round off the measurements to the nearest unit.

5   Mark the location of each object on your sheet of graph paper using the grid co-ordinates.

6   Measure and record the grid distances of each object in the room relative to the others.

7   Find out how many sides each shape is made up of. Each side is a vector.

8   Shade in only the whole grid cells (rasterising) which make up each object using different colours for each.

Chapter 4

GIS in schools

Infrastructure, methodology and role

Charlie Fitzpatrick and David J. Maguire

Geographical information systems (GIS) consist of robust hardware, software, data, and a thinking operator. Together, they provide powerful tools for mapping and analysing information about people, places and the environment. Recent improvements in computer hardware and software allow the powers of GIS to move effectively and affordably into the schools arena. With only a modest investment schools can have a new, full-featured GIS appropriate for classroom needs.

GIS can be incorporated easily into current curricula at all levels and subjects, supporting and enhancing existing activities instead of requiring an isolated, dedicated place. Teachers can teach with GIS as well as about GIS. The powerful tools permit teachers and students to explore and analyse information in a new way students' activities on the higher order thinking skills of observation, exploration, questioning, speculation, analysis, and evaluation.

In the dreams of educators, administrators and parents alike, school is a magical place. Students go to school and are engaged in stimulating, challenging and energising activities through which they learn about the world of today, how it came to be as it is, and how to make it better in the future. They learn to think both independently and critically, taking advantage of information from others, yet testing the ideas and casting them in new configurations. They learn about a multitude of topics, yet integrate the important ideas and information across disciplines. They develop skills and attitudes in one subject that transfer to all others. Such skills are critical for their present and future lives, and for the lives of all people around them.

In such a scenario, school is not without conflict or contest. Indeed, wrestling with ideas and coming to grips with different information, skills and values brings a steady stream of issues to explore. The challenge for educators is to lead participants toward exploration, integration and co-operation, instead of segmentation, stratification and conquest. GIS has the potential to bring all schools much closer to this ideal.

Defining GIS

A Geographical Information System (GIS) is a combination of elements designed to store, retrieve, manipulate and display geographical data – information about people, places and the environment. It is a package consisting of four basic parts: hardware, software, data and a thinking operator. The hardware is a robust computer. The software is a powerful set of instructions and procedures that can be used to solve problems. The geographical data are computerised information in a variety of formats. The thinking operator is a person who does not know all the answers, and may not even know all the questions, but who wants to learn about the cultural and natural environment and knows how to use tools creatively to look for patterns and processes.

Like any system, a GIS works best (perhaps 'only') when all parts are operating in concert. Again, like any system, the whole is far greater than the sum of its parts. To help explain what a GIS is and does, consider the analogy of an overhead projector.

Imagine an overhead projector, with a series of transparencies laid upon it. Each transparency is about your town, drawn to the same scale, and can therefore be integrated with the others. Each transparency deals with a different topic: rivers, roads, railroads, elevation, vegetation, land-use patterns, buildings, population characteristics, crime sites, churches and temples, school attendance boundaries, postal codes, utility lines and newspaper boxes. Standing before the overhead, you mix and match the layers at will, change classification schemes and modify symbols and colour combinations. You can zoom in and out, seeing all the information available or only the data you specify, comparing this layer with that feature, exploring the data in every way imaginable. As you play with these layers of information, relationships appear. This is in many ways similar to GIS. Through the power of a computer and software, using a wide range of electronic data, and with an eye toward patterns and relationships, GIS users explore geographical information. Through creative questioning, careful analysis, and even random exploration, GIS users learn the patterns of people, objects, and features of one site, how they interact, and how one region influences another. In short, GIS is a tool for learning about the world and all that is in it.

Hardware

The remarkable explosion of power in personal computers has presented enormous opportunities for the public. GIS used to be available only to those with access to the huge mainframes and powerful workstations necessary to churn large amounts of data. The recent increases in computer power, coupled with a steady decline in price, has made high-performance computers affordable for schools, even for those facing tight budgets. While configurations of hardware can vary widely, typical stand-alone personal computers suitable for GIS work in the classroom are likely to include:

- Pentium class or higher PC running Windows 95, 98, NT4, or higher
- minimum of 32 MB physical RAM (preferably more)
- minimum of 100 MB hard drive space (preferably much more) to dedicate to GIS
- Monitor with at least 256 colours (preferably more) and 640 × 480 pixels (preferably more)
- mouse
- CD-ROM drive (preferably fast)
- printer (preferably colour)

Going below these levels for a single-station setting in a classroom is not generally practical, because of the time required to accomplish tasks on less powerful machines. In an environment where ten seconds of delay can mean a significant drop in excitement, attention and productivity, speed is a critical element.

Software

Computer software is the instruction set that tells the hardware to accomplish tasks. Powerful software can use the rising capabilities of new hardware to its fullest extent, but the software must be intelligible for the user. Fortunately, however, the vast majority of tasks that schools wish to accomplish can be handled with a reasonable number of basic operations. Students and teachers generally need just the basic features of the software, and should not be concerned with learning immediately 'all there is' (a summary of basic tasks is presented below). More important than a thorough knowledge of the entire tool kit is a disposition for exploration and a capacity to think geographically – to search for relationships and patterns.

Data

Without data there is little point in having a GIS. There are several significant formats of data for even the novice GIS user to consider. First are the geographical data that represent the physical places: points, lines and areas. These data sets form the outlines and locations of places and features on a map. On a world map of population, for instance, one might expect to find a line layer for latitude and longitude, an area (or 'polygon') layer colour-coded by population, and a point layer for cities with the size of the points proportional to the population. The map might also have a line layer showing rivers. These geographical data sets show where something is on the planet and provide each feature a with unique name or an identifying number.

Next is the set of 'attribute data', or characteristics of features. These data sets are tables, made up of rows (or 'records') and columns (or 'fields') of information. In a 'countries' data table, for instance, each country would be

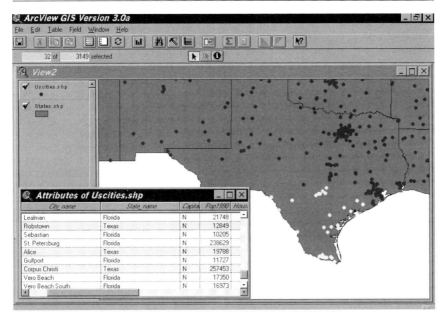

Figure 4.1 Screenshot of a typical GIS (ArcView 3) showing a map (US states) and attributes (names of US cities)

one record, and each piece of information about that country would go in its own field. Very quickly, one could have many pieces of information about many countries. There can even be many tables of country information. Matching each unique name in a table with the corresponding name in the geographical data converts dull tables into visually powerful displays. Countries are no longer uniformly coloured but, instead, shaded according to, say, population growth rate or density. Rivers are no longer single blue lines but can be graduated to show water volume or the amount of barge traffic upon them. Cities are no longer identical dots but sized and shaped and coloured to demonstrate one attribute or another. Geographical data and tables form the warp and weft of a GIS. By matching data tables to their counterparts in geographical data layers, GIS users in schools can map an endless array of information.

Operator

By this point, it should be clear why the most crucial element of the entire system is the operator. There are so many possible combinations for so many data items that the GIS needs careful guidance. Someone needs to find the most appropriate data. Someone needs to decide if – and why – one representation is better than another. Someone needs to identify the impact of seeing the data one way instead of another. Because a given set of data can be shown in many different ways, someone needs to determine what is accurate, appropriate, and complete – why the data are important and what message is transmitted by a particular configuration. During this exploration, important gaps in the data may be discovered. It may turn out that an existing data set has a hole or a flaw that needs to be fixed. The need for a certain set of additional information may become apparent only after significant tinkering has already occurred. Using GIS is an evolutionary process, involving exploration of what exists, wondering and hypothesising, evaluating results based on available data, identifying what other data are needed, and presenting interpretations based on the data used and the judgements made.

It is this last piece that points so clearly at schools. To be a positive force in society, to make well-considered decisions, to be a resource with the potential to improve the quality of life – people need to be comfortable with exploring and integrating information, seeking relationships, thinking critically, acknowledging differences, and finding common ground. Using GIS, these essential skills can be developed in schools, in all subject matters, by students young and old, and with all degrees of innate ability. They are indeed the very heart of GIS. The most delightful part about all of this is that using GIS is fun.

Incorporating GIS into schools

Levels

Teachers can use GIS tools successfully in all levels of school. However, it is important that teachers carefully match the task or opportunity with the developmental level of the students. A mismatch between student, educational tool, learning objectives and instructional method can render even the most powerful tool ineffective. No tool, by itself, can match the direction and customisation that a tuned-in teacher can give. While activities may vary widely between levels, the skills and principles endure. In each setting, exploration and discovery are key, with critical thinking skills to be fostered throughout.

Primary schools

In the early years teachers can introduce students to maps as models and geography as a discipline. Important concepts of absolute and relative location, place, region, scale, symbolisation and generalisation can be introduced. Students can begin to explore significant features of human and physical geography (mountains, cities, deltas, farmland, etc.) through, for example, satellite imagery. They can use local area maps of relevance (neighbourhoods, local watersheds, forests, etc.), perhaps produced by local GIS users. From such activities students should see that computers are tools with capacity for displaying information effectively, and that maps are dynamic and representational (thus distortable) displays rather than static and 'correct' portraits. Perhaps above all, students in the early years need to have fun exploring information on the computer; for this aspect, the multimedia linkage capabilities of modern GIS software should be exploited.

Middle schools

In middle schools, students can begin examination of special topics and regions with GIS. They can study a given phenomenon over space, seeing how it relates to others. They can survey the characteristics and relationships of geographically varied traits (population, economic makeup, physical features, etc.). They can examine a region and its features, and begin to understand the complex traits that identify and unite or separate areas, at scales local to global. Middle school is also the best opportunity for engaging in cross-disciplinary studies.

Secondary schools

Secondary school students can use GIS to expand the study of topics and regions, and explore geography-related job markets. Earlier regional and topical studies can be carried deeper or integrated with other studies. Students can focus on interrelationships between features and factors in other places. They can survey vocations that rely on the ability to gather, process, analyse and report spatially varied data. Guidance counsellors and teachers can explore the jobs that rely on spatial data (urban planners, marketers, environmental engineers, energy and utility companies, etc.) and how they gather, use and display their data.

Subjects

Teachers of geography are not the only ones with the potential to benefit from using GIS. Just as the tool can have powerful application in varied age ranges, so teachers in different subjects can explore a universe of topics. Once again, exploration and discovery are key, with critical thinking skills to be fostered throughout. In maths classes, students can use GIS to explore various map projections. They can investigate the process of locating objects with reference to relative and absolute markers. They can also use the rich spreadsheet functions and database graphing and charting capacities to study the mathematical relationships between factors found at various places.

Science classes can expand analyses of environmental relationships by displaying micro and macro systems as they occur over space. Local environmental analyses can be enriched by seeking similar patterns in other places. Students can explore the impact on visible patterns by altering the electromagnetic radiation received in satellite imagery. They can overlay these images with ground mapping to examine how people and the environment influence each other.

In language arts classes, students can enrich reading exercises by exploring the nature of a place using GIS maps. Conversely, geographical exercises can be supplemented by describing in writing the nature of a place, or the analysis of relationships. Art classes can be an especially important venue for GIS. Since the same data can be presented in many different ways and since a crucial aspect of mapping is how the untrained viewer interprets colours, symbols and patterns, the impact of various schemes warrants exploration. Because modem GIS can be a powerful multimedia tool – integrating still pictures, moving images, sound files, text files and external programs – art students can explore the richness of human expression with GIS.

Social studies classes, of course, can practically live in GIS, exploring decisions made in the past and how those decisions were shaped by differences in characteristics from place to place. With data as fresh as the daily news, students can explore current events, conditions, issues and conflicts, and see how their contributing factors are influenced by changes over space. All users can ponder the impact on societies caused by fundamental changes in technology and the types and amounts of information available.

Astute teachers will see the capacity for GIS as an integrative tool. Interdisciplinary studies are enhanced enormously when the same tool, skill, or concept finds a home in multiple settings. Because it is so all encompassing, GIS actually has the potential to revise the nature of education in profound ways.

With GIS, students can learn how the many diverse facets of the world are integrated. They can learn the value of exploration, wondering and questioning. They can see a single set of information being portrayed in many ways. They can develop their analytical skills, exercise integrative thinking and practice expressing their ideas effectively to others. They can learn about the richness of the world and all that is in it.

Methodology

The challenges for bringing GIS into the classroom invariably focus on 'How do we do this?' As happens with most new technologies, many potential users see single definite concrete answers when, in fact, the options are extremely varied. Still, there are some alternatives that teachers can consider – guidelines for thinking about the use of GIS in the classroom. In each case, the expectations need to match the possibilities.

One-computer classroom

Having only a single full-power GIS station for a class of twenty to thirty-five students can be a manageable setting that still takes advantage of the capacities of GIS. Two basic options are:

• Either project the display using an overhead projection panel or LCD projector, or alternatively use a large monitor. In this setting activities would be led by the teacher with all students working on a similar task. Focus in this setting can be as much about the technology as about the information. Students could, for instance, learn the basic operation of the software through a live group demonstration, or see modelled the 'explore-question-analyse-evaluate' process of geographical inquiry.
• Use a single full-power station to prepare a series of electronic or hard-copy output files. In this mode, students could work individually, in groups, or as a whole class. Here, the focus is weighted more heavily on working with the information, rather than on the technology.

Working in groups

Putting two or more students to a computer can increase learning, as long as the methodology and tasks are sound. Students of all ages seem to pick up most about the chosen subject when there are two pupils per computer. More than two per computer dramatically reduces the amount of time each person has for hands-on interaction, which limits students' total focus.

Group projects involving GIS could be organised to engage students with the technology, perhaps on a rotating basis. After some introductory training, and with a carefully designed set of activities, students can incorporate GIS activities with tasks at other stations in the classroom. For instance, in an examination of the school neighbourhood, different stations around the room might involve scrutinising photographs, studying historic documents, comparing data tables and translating mental maps to hard copy, along with working on the GIS.

Working solo

Individual work on GIS can be extremely powerful. Students can explore in depth in their own fashion and develop their own view of a situation. This can also result in widely different projects, 'study directions', results and timetables. Because of this, teachers should be extra careful to have realistic expectations. With extended training beforehand and appropriate coaching during the process, students can stay on a fairly consistent and very productive course. For this reason, teachers working with more than a dozen students at a time should view such solo efforts as projects for later in the school year, after developing requisite concepts and skills. When engaged in solo projects, students will have widely varied interests, approaches and results, even when working with consistent data, so it is extremely important to allow the opportunity to share projects with each other.

Network settings

Working with a computer network poses a special set of challenges and opportunities. Teachers in a school with a powerful local area network are in an excellent position to take advantage of GIS. However, computer network technology is tricky and requires special technical skills not common among subject- or age-focused teachers.

Therefore, working with GIS on a network should be explored carefully, well in advance of the anticipated event. GIS demands extensive computing power and often relies on large data sets, so the school's network 'traffic' may be much higher in GIS projects than for other network-based activities. The opportunities to engage in wide-ranging group projects, however, make networks an extremely attractive option.

Effective use of geographic information systems (GIS) involves more than just clicking buttons to create a map. In a school setting it means engaging in active learning, with significant thinking required. GIS can mean more to education than just 'having a source of maps'. GIS can affect the whole educational experience, for students, teachers and the community.

The role of GIS school curricula

The potential roles of GIS in schools curricula are many and varied. This section discusses some of the major areas in which GIS can contribute to students' educational experience.

Critical thinking

One of the major areas in which GIS can contribute to student's educational experience is through the development of faculties for critical thinking. In particular it can help develop the ability to analyse, synthesise and evaluate. In so doing, students' logical-mathematical, linguistic, spatial and interpersonal intelligence can be enhanced.

Logical-mathematical intelligence includes numeracy (the ability to interpret and use numbers and numerical skills) and technological capacity (the ability to understand and use tools which facilitate acquisition, processing and information). Linguistic intelligence includes literacy (the ability to interpret and present information in word form) and graphicacy (the ability to read and use visible symbols). Spatial intelligence, so important for thinking geographically, includes map literacy (the ability to transform real life into a mental or visual picture, or vice versa, at multiple scales). Lastly, interpersonal intelligence focuses on communication (the ability to transfer effectively to others, through multiple modes of representation, the information and knowledge gleaned through investigative processes).

Vocational tool

As well as developing critical thinking, GIS can be used to teach vocational skills. It should be obvious from the earlier discussion that GIS can assist in basic computer literacy. It can also be effectively used to provide integrated training in the process of research, including data gathering and preparation, storage, analysis, and presentation. Additionally, GIS activities provide actual training for many careers in government (planning, transportation and environmental health departments are major users), universities (in research and teaching) and in private businesses (GIS is widely used by utilities, retailers, environmental consultancies and software developers). GIS is very much a tool for the twenty-first century.

Educational reform

Perhaps less obviously, but in the long term more significantly, GIS can play a role in educational reform. As a new technology and educational paradigm, GIS helps promote change and growth for skill development, classroom organisation, instructional methodology, curricular content and community participation, all at the same time.

It is important to realise that a GIS does not contain or present 'the' answer. People define answers according to the questions they ask and the parameters they establish. A GIS provides methods with which to explore alternative responses for specific problems and situations. Users still need to define what constitutes a satisfactory answer to their question. Critical thinking plays a primary role in using GIS effectively. Explorations thus involve profound challenge for learners.

With GIS both students and teachers can be active learners at the same time. By developing new skills and exploring new understandings of a variety of topics, teachers can model for students the process and value of lifelong learning.

Furthermore, because the computer is a powerful tool for exploring similar content through divergent paths, students engaged in GIS can progress in varied ways, in a style and at a pace more appropriate for their individual interests, strengths and needs. Active exploration with GIS can more easily match the multiple modes of information access which different students need, while still affording each the chance to contribute to group activities and providing a powerful opportunity for constructing individual visions of the world. When adopting GIS centrally, assessment of student progress, achievement or development can be accomplished in multiple fashions, in ways that are appropriate to the students' interests, the school curriculum and the community needs.

Using GIS can also help students and teachers become more involved as local community participants and global citizens. Partnering with other GIS users in the community enacts the 'community as classroom' concept. Students, schools and the community all benefit as each pays closer attention to the needs of the others.

Summary

What this all means is that GIS can be a powerful ally in the effort to enhance education. Students and teachers can work together to build a coherent framework for information about the world. The community can share in the process of providing educational experiences and can gain from intelligence provided by the students. The focus on collaboration between students, teachers, school and community can provide significant long-term benefits for all.

Individually, students can benefit from increased attention to their strengths and weaknesses. The potential to relate schoolwork with explorations from everyday life can add powerful connections for students constructing their own views of the world. Engaging GIS at multiple levels and disciplines can yield an uncommon synergy in a setting too often fractured.

This is the vision and expectation built from early explorations of GIS in schools. Because of the speed with which this technology has burst on the scene, these descriptions are not supported by exhaustive studies. Similarly, because of the complexity involved, it has not been proven beyond all doubt that students and teachers who engage GIS develop the desirable traits noted above where such traits did not exist before. Rather, this is the view of individuals who have gained sufficient background with both the technology and the challenges of elementary and secondary education. As with much of education, the results may not be clear for years. But, given the current anecdotal evidence, there is strong reason to believe these positive statements are true for all students.

Finally, as technology brings us all ever closer to each other, it seems only too obvious that there is a need to understand more fully the changing relationships between people, places and the planet. We share many parts of our existence and need to explore our common ground.

Chapter 5

Teaching with GIS in Ontario's secondary schools

Bob Sharpe and Angela Crechiolo Best

Electronic atlases, computer-assisted mapping, the Internet, and Geographic Information Systems (GIS) provide teachers with an expanding suite of exciting tools for teaching and learning about geography and environmental studies. Until recently, however, these so-called spatial information technologies (SIT) have been limited to the resources and students of post-secondary institutions (Rogerson, 1992a, 1992b). This situation is now changing in Ontario with numerous reforms to the secondary school curriculum. In 1991, the Ontario Ministry of Education adopted mandates to develop guidelines for the integration of computers into all areas of the secondary school curriculum (Ontario Ministry of Education, 1992).

This trend in education policy is supported by a growing North American literature on teaching SIT at the pre-collegiate level. Research initiatives and international symposia, such as those sponsored by the National Center for Geographic Information and Analysis (NCGIA), provide rich resources for teachers. Also, the major GIS software vendors, such as ESRI, IDRISI, TYDAC and others, have developed, and continue to develop, aids for teachers. These efforts have resulted in training workshops, tutorials, and teaching manuals to assist educators to develop skills in the technical operation of GIS.

As of 1996, the process of integrating computers into the secondary school curriculum has been well underway in Ontario. Teachers are frustrated, however, by the limited access to hardware and software, by the lack of pre-service and in-service training, and most notably by the paucity of appropriate teaching resources. In particular, educators lack clear guidelines that apply GIS functionality to the teaching of geographic concepts and skills at the pre-collegiate level.

This chapter outlines aspects of the history and current status of GIS in Ontario schools by describing various efforts to promote the use of GIS in the classroom. It then outlines research undertaken at Wilfrid Laurier University to foster GIS education continuity between post-secondary institutions and secondary schools. Finally, it proposes a model curriculum for the integration of GIS into the pre-collegiate geography classroom.

The status of GIS in Ontario secondary schools

The status of GIS in Ontario's secondary schools between 1993 and 1996 was assessed through information collected from education journals as well as personal correspondence with individuals involved in GIS education. This information indicates a three-stage progression from a growing GIS awareness among teachers, through the provision of pre-service and in-service training, to current efforts in the development of curriculum. Each of these stages is described below.

GIS awareness

Teachers are made aware of the potential of GIS in the classroom through both personal and public sources. Evidence of a growing awareness can be seen in the various public outreach initiatives of organisations such as the Ontario Association for Geographic and Environmental Education (OAGEE), the Ontario Ministry of Natural Resources, Ryerson Polytechnical Institute and Wilfrid Laurier University.

The OAGEE's involvement in heightening GIS awareness entails providing a venue for GIS-related publications, and the advertisement of GIS workshops and vendor products in their periodical, *The Monograph*. Articles on SIT in *The Monograph* can be categorised into two main streams: the first identifying the benefits of computer integration from a pedagogical standpoint, that is, making the learning process as much fun, as interesting and relevant as possible (Gartrell, 1996; Robinson and Thorton, 1995; Holdren, 1992; Oliver, 1992; Robinson, 1991, 1990; Chasmer, 1990; LeDrew, 1988; Knapp, 1987; Wobscall, 1984); the second identifying and cataloguing specific software programs and Internet addresses (Robinson and Thorton, 1995; Brown, 1994; Oliver, 1994; Waterhouse, 1993; Colby, 1987). Although the discussion is primarily at the awareness stage of development in *The Monograph*, it has become increasingly more comprehensive over the last decade, and is sure to become more detailed and practical with articles on traditional teaching topics such as the five fundamental themes of geography (Koegler, 1995) and cognitive skill development (Augustine, 1988).

The OAGEE also promotes new techniques and pedagogical approaches to computer use in the classroom at their annual meetings. At the 1995 annual meeting 25 percent of the sessions were dedicated to the application of computer technology in the classroom. The sessions ranged from oral presentations to hands-on applications. Each of these services provide secondary school educators with a valuable source of information and ideas surrounding the introduction and integration of GIS into the geography classroom.

The Ontario Ministry of Natural Resources (MNR) has also had an influential role in the development of GIS in the education sector. For example, in 1995 the Ministry sponsored a workshop to facilitate an external transfer of resources and ideas from the Ministry to teachers. Other initiatives undertaken by the MNR include the development of the Hitchhiker's Guide to GIS at MNR. This demonstration software describes some of the basics of GIS and its application to environmental and park management. Originally designed for an elementary school-aged audience attending the Ontario Science Centre during Wildlife Week in 1994, the software has since been circulated to interested Ontario School Boards (Maher, 1995).

GIS awareness is also achieved on a more direct level through small workshops provided to educators during their Professional Development Days in Ontario. Some more prominent examples of such outreach efforts are Ryerson Polytechnical Institute in Toronto and Wilfrid Laurier University in Waterloo.

At the Applied Technology Department at Ryerson Polytechnical Institute, Dr Doug Banting has organised several GIS workshops for teachers. These short workshops have been attended by teachers from various Ontario school boards and involve a series of seminars. Tile presentations cover a variety of topics including an overview of GIS concepts, geographic skills, careers in GIS, and the cultivation of a geographic perspective (Palladino, 1994, 1995).

The Department of Geography and Environmental Studies at Wilfrid Laurier University has a twenty-two-year history of bringing teachers and academics together at an annual Geography Teachers' Symposium. In this way, the geography faculty has been able to identify on an ongoing basis the emerging needs and interests of geography teachers in Ontario schools. Since the 1980s teachers have expressed a growing interest in the use of microcomputers in the classroom. In 1992 a three-hour GIS workshop was first incorporated into the program. Since that time, several workshops have demonstrated the use of GIS in teaching and learning about geography.

Through workshops, conferences and educational journals, geography teachers in Ontario have become aware of the value and potential of GIS in the classroom. Having been made aware, however, teachers are now in need of more information, training, and resources for the integration of GIS in the pre-collegiate classroom.

GIS in-service training

In-service training initiatives are efforts by secondary and post-secondary institutions to provide instruction in GIS to teachers. Short workshops and summer institutes are typical approaches to training (Palladino, 1994).

There are at least four key teacher-training initiatives in Ontario. These are based at Queen's University in Kingston, at Napanee District Secondary School in Napanee, at Sir Sanford Fleming College in Lindsay, and at Wilfrid Laurier University in Waterloo.

Dr Roland Tinline, in the Department of Geography at Queen's University in Kingston, Ontario, has been a proponent of GIS in the secondary school classroom. Most notably, in 1990, he initiated a pilot project for GIS inclusion at the secondary school level through a regional geography course offered at Napanee District Secondary School (NDSS) in south-eastern Ontario. The project was received enthusiastically by students and led to a major proposal to introduce GIS at the secondary school level. This proposal involved a series of partnerships: the Canadian Centre for GIS in Education, the Lennox and Addington School Board, Napanee District Secondary School, the Ontario Ministry of Education, Queen's University's Faculty of Education and the Queen's/IBM Geographical Information Systems Laboratory (Tinline et al., 1991).

Mark Oliver, the NDSS course instructor for the geographics course, became the liaison between Queen's University and the secondary school sector. Due to the successful operation of the geographics course, Oliver's expertise is in increasing demand. This expertise has been made available to his Ontario colleagues through a series of regional workshops and summer institutes. To date, Oliver has worked with over 125 teachers and five classes of students (Oliver, 1996).

At Sir Sanford Fleming College (SSFC) in Lindsay, Ontario, faculty members from the GIS Applications Specialist Program (GISAS) have started to become involved with GIS in-service initiatives. SSFC and Trent University in Peterborough have joined forces to provide students enrolled in geography and/or environmental and resource science with experience in GIS. In the third year of study at Trent University, students can enrol in an eight-month diploma program at SSFC specialising in GIS applications. The program has important teaching applications, especially for those who are also enrolled in the education program at Trent University. Besides receiving a strong theoretical and technical background in GIS, students can focus on secondary school applications.

Another approach to GIS in-services for teachers at SSFC is a GIS course geared to secondary school educators. This course runs ten weeks in length, for three hours each week starting in March in 1996. The faculty from the GISAS program provide fifteen secondary school educators with thirty hours of intensive hands-on training in IDRISI and ArcView2. The use of existing data sets derived from the GISAS program were in the past used to train the teachers in the main areas of data acquisition, input, manipulation, analysis and output. Training in the application of GIS technology has remained the focus of the course. This has been supplemented in part by instructor-based assistance in the development of data sets and assignments for classroom use. Similar training efforts are being undertaken at Wilfrid Laurier University (WLU) in Waterloo, Ontario. Since May 1995, a program of Geographic Information Systems Certificate courses has been offered by the Department of Geography and Environmental Studies and Continuing Education. The GIS Certificate program is an outgrowth of the workshops offered at the WLU

Geography Teachers' Symposium. This program was devised to provide teachers with a more detailed and comprehensive introduction to GIS for use in the secondary classroom.

The GIS certificate program consists of three intensive two-day courses. Each course involves twelve hours of intensive theory and hands-on experience in IDRISI for Windows. Teaching methods are varied: mini-lectures, demonstrations and discussions, with an emphasis on experiential learning.

IDRISI for Windows was chosen as the demonstration software over other commonly found SIT software packages in Ontario schools, namely ArcView2, SPANS® Map, AutoCAD and MapInfo™, for a variety of reasons. First, IDRISI was developed at Clark University primarily as educational rather than commercial software. The software is available to schools at a substantially reduced educational price. In addition, the Windows interface makes it very user-friendly. Second, raster data structures are easier for the novice to understand when compared to object-oriented, vector data structures with their relational tables and topological information. Finally, IDRISI is compatible with the software and hardware available in most schools. The software has a wide range of import and export capabilities, and has a full range of input capabilities that do not require sophisticated and expensive hardware such as a digitiser or a plotter.

Curriculum development

GIS curriculum development takes place both at an individual level through teacher and student initiatives, and at a institutional level as part of larger academic development projects. At this early stage it is premature to annotate and assess individual efforts in GIS curriculum development so the focus here is on institutional initiatives.

Currently, the only GIS curriculum guidelines widely available to teachers in Ontario are the Ministry of Education's geographics course. The geographics course focuses on 'spatial information' presented in a variety of formats. Units draw on topographic maps, air photos, remotely sensed images, statistical analysis, mapping and charting techniques, pattern and locational analysis, and presentation skills (Oliver, 1993: 20). The use of SIT in the geographics course has emerged largely from initiatives at Queen's and the Napanee School District under the direction of Tinline et al. (1991) and Oliver (1996). The Geographics course was developed for use in the specialisation years (i.e. grades ten through [OAC]) and in order to expose students to spatial information technologies such as computer-assisted drafting packages and GIS software. The rationale for the course touches upon two main themes: first, the need to manage increasing amounts of information, issues and problems relating to society and the environment and second, the increasing demand for GIS expertise in both the workplace and post-secondary education. Now in its fourth year of inception, the geographics course has been adopted in an increasing number of schools across Ontario.

The curriculum guideline for the geographics course is structured in terms of educational outcomes and theoretical content. Educational outcomes refer to indications of student achievement in attitude, knowledge and skills. Attitudinal outcomes pertain to the acquisition of information and issues of access, bias, misuse and different dimension of the phenomenon being represented (social, political, cultural, economic, etc.). Knowledge outcomes centre around the identification of spatially significant data, the familiarity with basic map-reading skills and principles, and the investigation of patterns and relationships in the natural/cultural environment. Skill outcomes revolve around analysis and interpretation.

The theoretical content of the geographics course is organised around five central concepts: spatial information, reality to abstraction, abstraction to reality, locations and distributions, and applications of geographic techniques. These conceptual organisers were designed to acquaint students with:

- spatially significant information/data
- common data sources, basic map reading skills and GIS principle
- interpreting different forms of graphic media
- different levels of technology to input, manipulate and store spatial information
- geographical analysis (Ontario Ministry of Education, 1988).

While the geographics course focuses on teaching about SIT and GIS techniques, GIS can also be used as a teaching tool to balance the content and delivery of other courses. Courses emphasising the spatial distribution and patterns of social, economic, demographic or cultural dimensions of society can be readily presented to, or analysed by, students through a spreadsheet/database management system linked to a thematic mapper. This type of geographic analysis would be appropriate for regional courses such as the Geography of Canada (GCA) and Regional Geography (GRE), and population studies in Environment and Economy (GCE), Human Geography (GHU), World Development (GWD), Europe and Asia (GEA) and World Issues (GWI).

Courses emphasising the dynamics of the physical and human environment are also appropriate for computer integration. A GIS handles the selection, collection, analysis and synthesis of different bio-physical and cultural characteristics present on the land, enabling students to both characterise land uses and predict/model human-environment interaction. This type of analysis would be appropriate in the Physical Geography (GPH), Geography and Environmental Studies (GES), and Urban Studies (GUR) courses.

GIS in geography at Wilfrid Laurier University

Research and development at Wilfrid Laurier University

Research on the integration of GIS into the secondary school geography curriculum has been ongoing for two years at Wilfrid Laurier University. In particular, the GIS Certificate Program has provided an opportunity to evaluate the status of GIS in secondary education and to develop and evaluate a model curriculum for the integration of GIS into the geography classroom.

Since the program's inception in 1995, 136 participants have participated in six GIS courses consisting of four level-one courses and two level-two courses. These teachers represent approximately fifty schools from across southern Ontario. All participants shared an interest in learning about GIS, but they varied widely in their prior experience with microcomputer applications in geography. A third of the participants were computer literate and had previously used GIS, another third were comfortable with microcomputers but had no GIS experience and the remainder had little previous experience with microcomputers. Although program participants are self-selected and thus do not constitute a random sample, their views can be considered representative of the wider population of geography teachers in Ontario who are interested in GIS.

Within each course, time was allotted for an open discussion of issues relating to the integration of GIS into the curriculum. At the end, participants were asked to evaluate the course and its components. Further opportunities to test and evaluate different curriculum material arose during training sessions held at various high schools and in a focus group workshop held at the University. In addition, participants were surveyed regarding their access to computing resources and their intentions to teach with or about GIS.

Table 5.1 summarises the responses of the survey taken of program participants. Although the majority of secondary school geography departments have access to microcomputers, there is a wide disparity among schools. On average each school had access to thirty-one machines, although this ranged from a minimum of three in a small-town school to a maximum of 150 in a technical school in Toronto.

In the majority of schools the computers are located in laboratories that are networked to centralised servers. A smaller number of schools have stand-alone personal computer banks within their departments. Teachers noted that one disadvantage of the networked environment was the competition among other departments for the use of laboratories. In the past geography has not been a priority user of computers and thus is at a disadvantage in gaining sufficient access to the laboratories and other computing resources (Becker, 1991). On the other hand, many teachers recognise that a networked system can be supportive of a more integrative school curriculum, such as the common

Table 5.1 The current status of GIS in Ontario classrooms

Departmental access to computers	89	%
Average number of computers	31	
Minimum number of computers	3	
Maximum number of computers	150	
Personal computers	68	%
Networked computers	84	%
Centralised	92	%
Access to computer-assisted drafting (CAD) packages	43	%
Would jointly offer GIS with technical department	25	%
Would jointly offer GIS with other departments	58	%
Access to hardware		
Global Positioning System	7	%
Digitising tablet	30	%
Optical scanner	44	%
Monochrome printer	67	%
Colour printer	48	%
Plotter	30	%
Currently using GIS in the classroom	0	
Intend to implement GIS into the classroom	100	%
Geographics course	54	%
Within existing curriculum guidelines	56	%
Intend a software-driven approach	7	%
Intend a subject-driven approach	60	%
Intend a software and subject-driven approach	33	%

Source: GIS Certificate Program Wilfrid Laurier University 1995–96

curriculum in Ontario. Over half of the teachers surveyed said they would jointly offer GIS with other departments. They share the hope that GIS could be the focus of new multidisciplinary classes that integrate aspects of geography with environmental studies, physical and life sciences, computers, history, maths, government, vocational arts and economics (Palladino, 1994).

Access to hardware peripherals also varies widely among schools. In terms of data input devices, only a third of the teachers stated that they had access to digitising tablets or limited access to the tablets in CAD labs. More schools, however, have scanners or have access to commercial scanning. Only a handful of teachers in Ontario were using GIS in the classroom in 1995–6. However, fully 100 percent of the teachers surveyed intend to implement GIS into their teaching. Only 54 percent intended to offer the geographics course. This was due in part to budgetary constraints and an interest in not isolating GIS within one course, but to make it an integrative tool for all subject areas and courses. Teachers are hoping for the integration of GIS into curriculum guidelines of existing courses. Clearly their interest is to focus on teaching with GIS rather than teaching about GIS. The challenge then of curriculum design is to teach both concurrently.

In addition to the survey, focus group discussions revealed that teachers would like to see the following in the development of a GIS in geography curriculum:

- Usage of information resources that are readily available, that is, topographic maps and air photographs, and to an increasing extent, digital images from CD-ROM and the Internet
- Projects based on realistic data and issues
- Generic exercises that are not site-specific and thus allow teachers to adapt lessons to their individual locale
- A balance of GIS technical skills with geographical knowledge
- A progression from the most basic skills (i.e. mastering the Windows operating environment and IDRISI software specifications) and knowledge (i.e. both geographical and GIS-related) and progress to more expert levels (i.e. interpretation, analysis and modelling)
- The incorporation of stimulating and fun capabilities of GIS into the learning process (e.g. data visualisation and land-use change scenarios):
- Modular organisational with the flexibility to fit into a variety of time schedules. Ideally, curriculum units could be accomplished within the standard seventy-five-minutes class
- Transferability into different courses, that is, as a complete semester serving the geographics course, or as supplementary units within existing courses (e.g., physical geography, world issues, Canada)
- Detailed step-by-step instructions and comprehensive exercises
- Operability on basic microcomputers without digitising tablets

Based on these recommendations a model curriculum for the integration of GIS into the geography curriculum is being developed at Wilfrid Laurier University. The basic structure of the curriculum is described in the next section.

GIS in geography: a model curriculum

The model curriculum is organised as series of progressively more challenging levels. Students work from the novice towards expert levels in terms of both GIS functionality and geographic knowledge. Table 5.2 illustrates how GIS can address specific geographical questions from the novice or low level of difficulty to the expert or most complex level of difficulty.

The framework described on Table 5.2 is generic and does not apply to the GIS functionality of specific software. Rather, the focus is on the types of questions a geographer might ask and how, in a general context, GIS can address these enquiries. Additional columns can be added to the framework showing the IDRISI or other software commands appropriate to each type of geographic question.

Table 5.2 A general approach to teaching geography with GIS

	Questions asked	*Geographic theme/ concept*	*GIS functionality*
Low	What is where? Queries on individual elements, primarily descriptive in nature	The identification of individual elements that together form the building blocks for specific themes, concepts, principles or ideas	Graphic display, map composition and manipulation to clarify features and enable simple queries on individual image elements
Medium	Where are what? Queries on two or more variables high-lighting associations, patterns or relation-ships.	The identification of relationships, associations and patterns that evolve out of the interaction between individual sets of elements	Data manipulation and analysis through data base query (SQL) and reclassification and Boolean, logical and/or mathematical overlays resulting in the identi-fication of structural relationships that exist between two or more static elements
High	What if? Modelling dynamic relationships between two or more variables	The emergence of key concepts, themes, principles or ideas	Modelling and predicting dynamic relationships between two or more variables; visualising uncertainty; decision support systems; time series analysis; spatial statistical analysis

To implement this framework a series of exercises were devised based on the application of IDRISI to problems of geographical enquiry and the representation of physical and urban landscapes. These exercises were designed to meet the needs of geography educators in terms of both GIS skills and geographic content. The exercises are organised in three levels: A, B, and C.

Level A consists of two exercises that familiarise the novice with IDRISI for Windows and introduce basic GIS concepts. Both exercises are self-directed and provide detailed step-by-step instructions. Exercise 1 systematically demonstrates the functionality of key menu items and icons of the GIS software. This is most effectively done as a demonstration by the instructor using the sample data sets that come with the software or with home-grown examples. Tutorial 1 in the IDRISI manual is a good model for this exercise. Exercise 2 introduces some basic GIS concepts and instructs how spatial and non-spatial data are input, manipulated and represented within GIS. It is designed to be a stimulating and fun project in which students explore the visualisation powers of GIS by creating a simple three-dimensional surface model and by creating several different map compositions. At the end of the exercise students are able to generate basic maps that can identify 'what is where?'.

Subsequent exercises in the curriculum are designed as projects or case-studies in which GIS is applied to solve a particular geographic problem. Each project is best started with a class presentation by the instructor on the overall objectives, organisation and conceptual background of the project. Then students, working first individually and then in pairs, proceed through a series of self-directed exercises. Occasional presentations to the class might be needed to define and explain some of the geographic and GIS concepts used in the project. Other resources, including journals and the Internet, can be brought in to enhance the delivery of the conceptual background.

Level B consists of two projects that differ in substantive focus. One project emphasises aspects of physical geography by further exploring the properties of a three-dimensional landscape model. Through a series of exercises students derive spot heights from a topographic map, georeference an airphoto and integrate the two sources. The second project emphasises aspects of human geography by examining land-use change in a rural-urban fringe. A series of exercises grade students in the creation and analysis of raster and vector data extracted from a time series of topographic maps. Both these projects go beyond the basic queries of level A and involve analyses that address the question 'where are what'?

Level C includes a variety of possible projects that build on the geographic and GIS concepts introduced in previous levels. For example, one project uses the same air photograph from level B to interpret and classify forest cover. Students can then examine aspects of the relationship between forest cover as derived from the airphoto and topography as derived from the topographic map. In another project students are given the problem of selecting the best site for a waste disposal facility in the rural-urban fringe given their own, or specified, locational criteria.

Students wishing to go beyond level C might be encouraged to develop a project of their own that includes some data collection, manipulation, analysis and output. Alternatively, teachers might co-ordinate students in the development of larger-scale and longer-term projects of relevance to geographical-environmental concerns in their local community. For example, students might develop an environmental information system for their local watershed (Sharpe, 1995). Or, students might provide GIS services to local companies and even the local school board. Projects at levels C and higher challenge students to use their news skills in geographic information in solving more complex problems and modelling human-environment relationships.

Teachers can devise a wide range of evaluation methods to assess the student's progress through these levels. Listed below are just a few evaluation methods that complement the proposed model curriculum:

- Map compositions at the end of each module and every project
- Log books or records for tracking and managing files
- A portfolio or folder for saving map compositions and reports in progress
- A final report outlining the methods used and presenting the key findings
- Short answers to questions located throughout the exercises
- Quizzes and tests of both concepts and skills
- A student-generated glossary of concepts and terms

Conclusions

In Ontario there is widespread and growing awareness among geography teachers about the potential of GIS as a useful educational tool. Teachers realise that it is increasingly difficult to motivate geography students without using microcomputers, SIT, and GIS. A challenge to educators, however, is to teach GIS as a tool without losing sight of the broader objective of learning about geographic concepts and skills.

This chapter argues that an immediate obstacle to the further integration of GIS into the geography curriculum is the lack of appropriate curriculum guidelines, teaching resources, and instructional manuals. Responding to that need, a model curriculum was proposed that links the teaching of GIS with the teaching of geography. This curriculum now needs further development. In particular, it is important to find ways of emphasising instruction in geography while making the GIS technique as transparent as possible. Additional projects and teaching resources are needed that focus on a wide range of geographic applications. For this purpose it would be valuable for educators to meet on a regular basis to share their GIS teaching applications and experiences. Finally, GIS software capabilities, such as macros, and customised interfaces and programs need to be more fully exploited to create GIS applications dedicated to teaching geography at increasing levels of complexity.

References

Augustine, H. (1988). 'Growth in Cognitive Thinking Skills', *The Monograph*, Vol. 39(3), pp. 11–15.
Becker, T.W. (1991). 'How Computers are Used in the United States Schools: Basic Data from the 1989 1.E.A. Computers in Education Survey', *Journal of Education Computing Research*, Vol. 7(4), pp. 385–406.
Brown, M.A. (1994). 'Shareware: An Alternate Source of Geography Computer Programs', *The Monograph*, Vol. 45(2), pp. 7–8.
Chasmer, R. (1990). 'Computer Experiences in the Geography Classroom', *The Monograph*, Vol. 4(2), pp. 14–16.
Colby, D. (1987). 'Computer Applications for Geography Classes', *The Monograph*, Vol. 38(1), p. 9.
Gartrell, J. (1996). 'Geographic Information Systems in the Classroom: Geography at West Hill Secondary School with GIS and more ...', *The Monograph*, Vol. 47(1), pp. 21–2.

Holdren, T. (1992). 'Geographic Information Systems in Geography', *The Monograph*, Vol. 43(3), p. 17.

Knapp, K. (1987). 'Making Geography a Today Subject in Ontario Schools', *The Monograph*, Vol. 38(1), p. 10.

Koegler, D. (1995). 'Exploring the Five Themes of Geography', *The Monograph*, Vol. 46(3), pp. 6–10.

LeDrew. E.F., 1988. 'Microcomputer Based Remote Sensing in the Geography Curriculum', *The Monograph*, Vol. 39(1), pp. 11–19.

Maher, B. (1995). 'The Virtual Landscape: Resource Information Management at the MNR', Keynote Address: Access Workshop, May 27.

Oliver, M. (1992). 'Computers in your Geography Program', *The Monograph,* Vol. 43(3), p. 16.

Oliver, M. (1993). 'Geography Meanders into the 21st Century – early', *The Monograph*, Vol. 44(3), pp. 20–1.

Oliver, M. (1994). A Summer Institute for Geography Educators, *The Monograph*, Vol. 45(2), pp. 14–15.

Oliver, M. (1996). Personal correspondence, October 1996.

Ontario Ministry of Education (1992). *Integration of Computers Across the Curriculum*, Ontario Ministry of Education.

Palladino, S.D. (1994). 'A Role for Geographic Information Systems in the Secondary Schools: An Assessment of the Current Status and Future Possibilities', Unpublished Masters Thesis, Department of Geography, University of California, Santa Barbara.

Palladino, S.D. (1995). Personal Correspondence, July 12.

Robinson, C. (1990). 'Computer Bits', *The Monograph*, Vol. 41(4), pp. 4–5.

Robinson, C. (1991). 'Computer Bits', *The Monograph*, Vol. 42(2), pp. 27–8.

Robinson, O. and Thornton, B. (1995). 'Critical Thinking and Computers: A Natural Combination', *The Monograph*, Vol. 46(3), pp. 17–21.

Rogerson, R.J. (1992a). 'GIS and Canadian Educators', in *Geographical Information 1992/3: The Yearbook of the Association for Geographic Information* (Eds J. Cadoux-Hudson and D.I. Heywood) pp. 45–50, London: Taylor & Francis.

Rogerson, R.J. (1992b). 'Teaching GIS using Commercial Software', *International Journal of Geographic Information Systems*, Vol. (4), pp. 321–31.

Sharpe. B. (1995). 'Environmental Education and Environmental Information Systems: A Project on the Grand River Basin, Ontario', *International Research in the Journal of Environmental Education*. Vol. 4(1), pp. 101–6.

Tinline, R., Alderson, C. and Mansfield, D. (1991). 'Proposal JOP pilot project: Geographic Information Systems in the Secondary School Curriculum'.

Waterhouse, T. (1993). 'Simcoe: A Geographic Information System', *The Monograph*, Vol. 44(3), pp. 24–5.

Wobscall, S. (1984). 'The Use of Computers in Teaching Geography', *The Monograph*, Vol. 35(1), pp. 2–4.

Further reading and information

Green, D.R. (1992). 'GIS Education and Training: Developing Educational Progression and Continuity for the Future', in *Geographic Information 1992/3: The Yearbook of the Association for Geographic Information* (Eds J. Cadoux-Hudson and D.I. Heywood) pp. 283–95, London: Taylor & Francis.

Green, D.R. (1993). 'Introduction: GIS Education and Training – Progress in 1993', in *Geographic Information: 1994. The Yearbook of the Association for Geographic Information* (Eds D.R. Green, D. Rix and J. Cadoux-Hudson) pp. 329–31, London: Taylor & Francis

Green, D.R. and McEwen, L.J. (1993). 'The User-Friendly Interface: An Essential for GIS Education and Training', in *Geographic Information 1994: The Yearbook of the Association for Geographic Information* (Eds D.R. Green., D. Rix and J. Cadoux-Hudson) pp. 355–63, London: Taylor & Francis.

Kemp, K.K.., Goodchild, M.F. and Dobson, R.F. (1992). 'Teaching GIS in Geography', *Professional Geographer,* Vol. 44(2), pp. 181–90.

Marran, I.F. (1994). 'Discovering Innovative Curricular Models for School Geography', Vol. 93(1), pp. 7–10.

OAGEE (1993). 'Geography: A School Subject for the Past, the Present, and the Future', Executive Summary, *The Monograph*, Vol. 44(3), pp. 7–12.

Ontario Ministry of Education (1988). *Geographics Geography Intermediate and Secondary Education*, Part E, Ontario Ministry of Education.

Ontario Ministry of Natural Resources (1995). *The Hitch-Hikers Guide to Information Sheet*, Ontario Ministry of Natural Resources.

Palladino, S.D. (1992a). 'GIS and Secondary Education in the United States', in *Geographic Information 1992/3: The Yearbook of the Association for Geographic Information*, (Eds J. Cadoux-Hudson and D.I. Heywood) pp. 304–9, London: Taylor & Francis.

Palladino, S.D. (1992b). 'GIS in the Schools', Workshop Resource Packet, *Technical Report 93-2*, National Centre for Geographic Information and Analysis, Santa Barbara, CA.

Palladino, S.D. (1993). 'An Update on GIS in US Secondary Schools and NCGIA Secondary Education Project Activities', in *Geographic Information 1994: The Yearbook of the Association for Geographic Information* (Eds D.R. Green, D. Rix, and J. Cadoux-Hudson) pp. 340–45, London: Taylor & Francis.

Tinline, R. (1995). 'Geography in the 21st Century', Keynote Address: Access Workshop, May 28.

Unwin, D.J. (1991a). 'Using Computers to Help Students Learn: Computer Assisted Learning in Geography', *Area*, Vol. 21(1), pp. 30–312.

Unwin, D.J. (1991b). 'GIS in the Curriculum', in *Geographic Information 1991: The Yearbook of the Association of Geographic Information* (Eds J. Cadoux-Hudson and D.I. Heywood) pp. 37–42, London: Taylor & Francis.

Unwin, D.J. (1992). 'Computer Assisted Learning in Geography', *Computing and Education*, Vol. 19(2), pp. 73–8.

Walsh, S.J. (1988). 'Geographic Information Systems: An Instructional Tool for Earth Science Educators', *Journal of Geography*, Vol. 89(1), pp. 17–25.

White, K.L. and Simins, M. (1991). 'Geographic Information Systems as an Educational Tool', *Journal of Geography*. Vol. 92(2), pp. 80–5.

Wood, A. and Cassettari, S. (1992). 'GIS and Secondary Education', in *Geographic Information 1992/3: The Yearbook of the Association for Geographic Information* (Eds J. Cadoux-Hudson and D.I. Heywood) pp. 296–303, London: Taylor & Francis.

Chapter 6

GIS in secondary school geography curricula

Mark Oliver

Background and rationale

In recent years, geography as a discipline in secondary schools has seemingly been classified as one that is not as important as the others. In the era of mathematics, science and technology, geography has fallen on hard times. In fact, the latest Royal Commission on Learning in the province of Ontario omitted geography as a requisite component of a person's education, substituting instead the generic term 'social studies'.

This downplaying of the discipline has not gone unnoticed by parents and students and administrators in the school system. Enrolments have been falling along with the prestige. One can almost hear the parental decries of 'that course won't get you a job'. When difficult staffing or budgetary decisions must be made in the system, they invariably fall against geography in favour of more popular contemporary fields of study.

All is not lost, however, as the recognition of the fact that they were 'under attack' has resulted in geographic educators developing a strategy for survival.

Traditional secondary school geography curricula include courses such as 'physical geography', 'world regional geography' and 'world human geography'. Whilst to a dedicated geographer these titles reveal that inside the courses there is an endless amount of interesting information waiting to be discovered, many teenage students appear unable to make that connection. Consequently the titles have been updated, for example, from 'physical geography' to 'natural disasters', from 'world regional geography' to 'travel and tourism' and from 'world human geography' to 'world issues'.

It has worked to a degree. Student enrolment in geography courses generally stopped its downward spiral trend and in many cases went up. Aggressive educators started using computer data sets to augment their classroom activities and to further the modernisation of their teaching.

The calendar of course offerings for secondary school geography still lacked an up-to-date course dealing with what geographers do best: namely, assessing a situation, collecting data for it, analysing the data, and graphically presenting the analysis. The coming of age of computers only enhanced this component of geography and further exacerbated the need for a course offering the associated GIS skills.

In 1988, the Ministry of Education for the province of Ontario (now the Ministry of Education and Training) approved a course entitled 'geographics'. This course focuses on topics like 'spatial information', 'reality to abstraction with data', 'abstraction to reality with data', 'locations and distributions' and 'applications'. Whilst encompassing the traditional geographic skills this course lends itself perfectly to the introduction of GIS as the approach to the utilisation of these skills. The 'geographics' course gained further credibility when the Ministry introduced its 'computers across the curriculum' policy.

Although the geographics course provided the opportunity to fill a void in the curriculum, few geography educators had experience with this approach to their discipline and so the course was virtually non-existent.

Corporate and educational partnerships

In 1992, a proposal to conduct a pilot project designed to introduce GIS into secondary school geography curricula (using the geographics guide) submitted to the Ministry jointly by Queen's University and the Lennox and Addington County Board of Education was approved and the process of developing a true 'high-tech' geography course began.

The project was based at Napanee District Secondary School, a rural school with a student population of approximately 1,500, located in Napanee, Lennox and Addington County, Ontario.

The Ministry provided funding for educational time release (enabling teachers to participate in this venture), training, software, travel and independent evaluation of the effort. The author served as Project Co-ordinator whilst the direction for the effort came from a 'working panel' including two others, namely Dr Roland Tinline, Queen's University GIS Laboratory/ Geography Department, and Dickson Mansfield, Queen's University Faculty of Education.

This 'panel' planned and assisted a 'teacher group' comprising fifteen geographic educators from five contiguous counties. This organisational structure was designed to utilise the expertise of the staff at the Queen's GIS laboratory while facilitating the dissemination of the knowledge to the Faculty of Education and to the teachers involved. Queen's GIS staff initially provided introductory training to the members of the 'teachers group' and facilitated visitations by teachers and students to the Queen's GIS lab.

By September 1992, a sixteen-station GIS laboratory was established at Napanee District Secondary School by the Lennox and Addington County Board of Education. One of the most important tasks was to select a GIS software package that could be used by secondary school students. Two major skills that the 'panel' considered critical for the students to acquire were those of digitising maps and performing fundamental spatial analysis.

The 'target' students were in the third year of secondary school and the assumption was made that these students would not be computer oriented. This mandated that the software chosen had to be very user-friendly while maintaining a certain level of power and flexibility.

At this time, the Canadian Centre for Studies in GIS, located in Ottawa, recommended the program SPANS® Map from TYDAC Technologies. This 'desktop mapping' program which incorporates spreadsheets, charts, maps, appended images and appended text was provided by TYDAC and used to introduce students to spatial analysis. TYDAC not only provided the software but also numerous data sets and technical support as well.

For the purpose of teaching students how to digitise maps, AutoCAD software from Autodesk was selected as the most appropriate tool. This program was selected for a variety of reasons, not the least of which included the fact that many schools already had AutoCAD software in their architectural drafting facilities and so the introduction of this new geography course would not become a financial burden for the educational institution. Autodesk Canada provided AutoCAD to Napanee District Secondary School to facilitate this aspect of the project.

Statistics Canada was in the process of developing a CD-ROM version of the Canadian Census and provided the program E-STAT as a database for student use.

Armed with these three tools, and very little personal expertise involving any of them (or computers for that matter), the Project Co-ordinator began the introduction of GIS as a course (geographics) in secondary school geography.

Impact on students and programme

The first group of students involved had, in fact, registered for a 'world human geography' course and not the geographics course because approval for the pilot project came after students had selected their courses for the ensuing school year. These sixteen students were given an explanation about how they would serve as 'guinea pigs' and were given the option of changing their course selection if they felt uncomfortable with this new course. None opted out.

After about a month, word had spread in the school about the 'neat' things going on in the new geography computer lab. Enrolment actually rose to twenty-four students as the term progressed.

The normal growing pains associated with computers were experienced by all involved but were surpassed by the learning experienced by one and all.

The most significant result of the introduction of GIS into geography has been in the area of student response to the programme. When this course was offered to the school the next year, over seventy-five students signed up. The subsequent year over 150 students enrolled and in September 1995, 170 students were registered in GIS-based courses at Napanee District Secondary School (in fact student demand has been so strong, that a second, upper-level course has been developed incorporating remote-sensing software from PCI, GPS work using a Magellan unit with differential correction, more sophisticated spatial analysis with SPANS® GIS and SPANS® Explorer and advanced data management with Autodesk's ADE package).

In spite of this rapid growth in student enrolment in the GIS courses, student participation in the traditional geography courses has not experienced any negative side effects. Rather, many students tend to be 'doubling-up' in their geography courses enrolling in the traditional courses like 'physical geography' while also taking the computer-driven geographics course.

There can be no denying the extremely positive effect on the students' self-esteem that has been derived through their use of the 'industrial strength' software. There seems to be an extremely suitable balance between the complexity and sophistication of the software and the students' abilities to complete projects successfully and professionally (it is truly rewarding to see a seventeen-year-old student want to take his/her work home to show it off). This manifests itself in student punctuality and attendance patterns that are unequalled elsewhere in the school. Indeed, geography is considered the 'high-tech' department at Napanee District Secondary School!

Teaching strategies

The students enrolling in the geographics course tend to be non-computer oriented. In other words, they likely have taken a 'keyboarding' class but may not have taken any courses in 'programming'.

In this course, after a brief introduction into spatial analysis, one of the software packages is selected as a starting point. The students are 'walked through' a series of orientation exercises following directions that are provided verbally, through a PC Viewer and on paper. At each progression, a simple exercise is given to the students that is designed to reinforce the learning that has been completed. Upon completion of the orientation exercises, students are assigned a larger project that involves their use of all of the skills previously introduced with that software package. An example with SPANS® Map might be to have the students generate a series of graphics depicting population density shifts over three decades for a region. For AutoCAD the exercise might be to digitise a township from a 1 : 50,000 topographic map. The skill development process is repeated for the other software that is available for the students to use. Students are then provided with the opportunity to design a project and to utilise the technology at hand to investigate the topic and to present their findings.

Community projects

Once the students have become reasonably proficient in their use of the technology a tremendous opportunity exists for the educator.

There is simply no end to the opportunities for mapping and graphics. Students' projects can be designed for the ability of the individuals and can range from mundane to extraordinary. Student work at Napanee District Secondary School has included such efforts as: an updated school map showing wheelchair access points, school bus route maps, maps for the Napanee Region Conservation Authority (see figure 6.1).

Each of these items has the logo of the NDSS GIS lab on it. In some cases, thousands of these maps were reproduced and distributed. This generates a great deal of good 'PR' and seems to help in the situations where some of those tough administrative decisions could go against geography. Schools thrive on positive press and GIS allows geographers to generate it.

Province-wide introduction of GIS

Although at the onset of the project the 'panel' and 'teachers group' believed that the introduction of GIS into the geography curricula could and would be successful, all involved were amazed by the degree of success.

Part of the mandate of the pilot project was to conduct workshops for educators in each of the six districts of the province as recognised by the Ministry. It was deemed most appropriate for educators to experience this GIS phenomenon first hand rather than through a slide show or overhead presentation. The difficulty with this concept was that no geography educators had access to a laboratory of computers that were powerful enough to run the software (at the time that meant a 386 with a math co-processor and 8

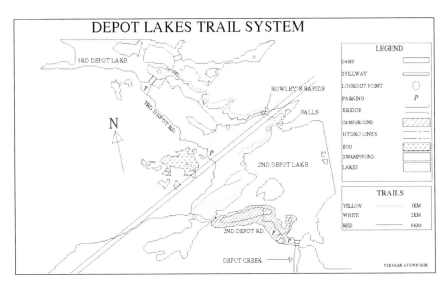

Figure 6.1   A map created for the Conservation Authority was selected by Autodesk Canada to be used in a national advertising campaign

MB ram). The solution was for the computers from the GIS lab at Napanee District Secondary School to travel to the sites of the workshops. It was decided that a ten-station portable laboratory, complete with digitising tablets and a CD-ROM unit, would be adequate for most workshops. During 1993 and 1994, workshops using SPANS® Map, AutoCAD and E-STAT were conducted in eleven cities for geography teachers and administrators from over one hundred secondary schools. Concurrently, sessions were conducted for Faculty of Education students at Queen's University in order to better prepare them for their future as geography teachers.

The primary focus of these workshops was to expose geography teachers to what may be considered as the future of geographic education and to spark an interest in them to learn more about this field. While successful in this regard, it became apparent quickly that teachers needed an opportunity to obtain training in this area without returning to university or college.

Teach the teacher

In response to the recognised need for training in GIS, a Summer Institute was held for the first time in 1994. This was a two-day, hands-on training session with one day devoted to SPANS® Map, one day for AutoCAD and an evening for E-STAT. It was held at Napanee District Secondary School during the first week of July. The session sold out and a second one was organised.

The 'Summer Institute in GIS' was repeated in July, 1995. Both sessions sold out again and this time the corporate sponsors, TYDAC, Autodesk and Statistics Canada played significant roles in bringing training to teachers by presenting at the sessions and by providing some software to participants.

The 'Summer Institute' attracts the most aggressive and adventurous geography teachers who genuinely believe that the GIS approach for their department makes sense. In the two summers about fifty-five teachers have been given the fundamentals they need in order to get started in their schools.

The sessions involve the same orientation walk-through that the students get at NDSS and the participants get paper copies of all the exercises and instructions. The teachers present go from how to use a 'spreadsheet' in SPANS® Map to how to set up the universal transverse mercator (UTM) co-ordinates for the tablet in AutoCAD. As well, they see examples of the students' work and get a chance to use some hardware like plotters, digitising tablets and GPS units before they have to make decisions about purchases for themselves. They also have the opportunity to peruse a curriculum toolkit (available from the author) that has been developed for teachers and to browse through the industry periodicals.

The two-day session is, as most people realise, not enough time to master much. There may be a need in the near future to incorporate a second session that will provide a more advanced set of skills for the teachers. The session as it now stands is, however, designed to provide participants with a successful set of experiences upon which they can build. Part of the province-wide implementation strategy included the teachers who attend the 'Summer Institute'. The plan was for those people to go back to their schools and try getting some GIS into their curricula. After some time, these teachers will feel comfortable sharing their expertise with their peers and the knowledge base and teaching base will broaden.

GIS at the secondary school level is still in its infancy. However, in 1992 there was only one school, NDSS, teaching these skills. Now there are dozens and more are coming! Currently Napanee District Secondary School has its second, more involved GIS based course, and others schools will surely follow this pattern.

As western society enters the information age, while simultaneously the environment is increasing in its global significance, it is imperative that students learn to analyse voluminous amounts of data using computer technology. There is no better discipline for the fulfilment of this goal than geography. Computerised spatial and temporal analysis belongs here!

Chapter 7

Geographical Information Systems

An introduction for students

Michael J. Brown

The Coastal Studies and Technology Center at Seaside High School in Seaside, Oregon, USA is a non-profit organisation which exists in a small (600) student 9–12-year-old high school in rural Northwest Oregon. The Center provides opportunities for students, staff, and community members to work together to investigate, research, and manage natural resource issues in our local and regional community. Students come to these projects at The Center by participating in high school science coursework at Seaside High School.

Technology, including Geographical Information Systems (GIS) and Global Positioning Systems (GPS), play an important part in being tools for the student staff to work with and to interact with scientists, community leaders, and citizens in their work. Students are trained in the basic fundamentals of these technologies and use them in many of their projects. On a yearly basis over one hundred students are trained in GIS fundamentals and become beginning users of the GIS program ArcView.

Our Center has a GIS partner, the Columbia River Estuary Study Taskforce (CREST), which gives us technical advice, shares data, and can help in digitising data points. Because Geographical Information Systems are complex and the concepts are hard to master at first, a curriculum pathway was designed to bring students conceptually through a series of simple mapping techniques towards the multilayer concepts of digital mapping. The main goal is not to teach GIS as an end to itself, but to use it as a tool to observe geographical relationships in a more powerful way. Students would then use that technology in their work with professionals to help organise data in a more meaningful way to gain greater insight about our research.

There are many ways to learn GIS and this is just one of them. It seems to make sense to high school students and as they build their skills they will understand the ideas and concepts that are inherent to ArcView and other GIS programs. Many organisations contributed to the making of our curriculum. Partnerships in creating a GIS program at any level is very important. More information about this curriculum or program can be found by writing to the author.

Geographical Information Systems

Geographical Information Systems (GIS) allow for the input, manipulation, and analysis of various forms of geographic data. It is a powerful technology that provides for a way to view complex amounts of information in the visual format of a map. This technology is currently being used by many types of businesses, industries, agencies and communities who are trying to analyse complex issues and find answers to them. Many types of data can be sorted, compared and connected through this technology. Available geographic information and high-power desktop computers have moved the availability of this technology into the realm of secondary schools. Students can and should have this information and the technological means for investigating the complex management issues surrounding natural resource planning in their own local communities. GIS integrates geographic information, computing, cartography and decision-making skills in looking at many types of information which can then be analysed via the GIS software and through which new insights and decisions can be made for understanding planning and resource issues.

This curriculum is designed as a quick way for students to start to understand some of the ways that GIS can become a useful tool for them. It will help them start to understand how different types of data can be compared and integrated as well as giving them some skill in working on real-world projects that are being done by scientists and agencies right now. The combination of GIS technology and students will benefit the overall science and social community.

GIS, as a tool, is currently being taught with a curriculum aimed at the post-secondary-level student. Several companies and many educators are working to bring this powerful tool to the K–12 (Kindergarten to Grade 12) arena for students to use to understand their world. This curriculum is a first effort to bring this technology into students' hands, give them some skills and understanding, and provide ideas for uses and projects which are occurring and need students to complete. As a first effort, this curriculum is by no means a complete work and draws upon many other people's work to point the way. There are also other efforts in curriculum and those should be sought out for additional insight into the possibilities.

What is GIS?

GIS information can be very diverse and can come from a variety of sources. It could be census information, road locations, forest boundaries, zoning maps, utility maps of water, sewer and electric lines, and lots of other types of data. These could even include orthographic photos or satellite imagery. Geographic information software and powerful computers can give you the ability to quickly integrate and organise that database around a database

> ### What is "GIS"?
>
> Geographic
> Information
> System
>
> GIS is a system designed for storing, analyzing,
> displaying, and manipulating spatial data –
> information about places

Figure 7.1 What is GIS? (Source: taken from ESRI ArcView Explorer; graphic image supplied courtesy of Environmental Systems Research, Inc. and is used herein with permission)

structure. This structure can then be looked upon as layers with each separate set of data forming its own layer. The technology gives you the ability to look at each layer, a combination of layers, or to search for a particular piece of data within a layer.

GIS data

In a GIS each category has its own layer. Each layer can have many different elements. The total system stores lots and lots of data.

The data that can be looked at can come from a variety of sources. The computer and the software allow relatively easy ability to control individual attributes of each layer for obtaining how you want the data displayed or what information you seek to see on the screen.

Spatial analysis: what is it and what does it mean in reference to geographical information

Spatial analysis is the process of:

- looking at what is where and where things are in relation to others
- integrating different types of data
- asking specific questions about how events, patterns, or conditions in one place affect those in other places
- using maps interactively, to display the data, to generate new questions, to provide new views of the situation and thus to produce better understanding

New questions, old questions, and the main question: what ought we to do?

Geographers try to figure out why things are where they are. They analyse the patterns of what is where and look for all the whys. All these events and

What sort of data is used in a GIS?

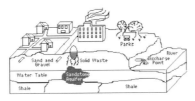

What kinds of data are represented here?

Figure 7.2 What sort of data is used in GIS? (Source: graphic image supplied courtesy of Environmental Systems Research, Inc. and is used herein with permission)

In a GIS, each category has its own "layer". Each layer can have many different elements. The total system stores lots and lots of data!

Figure 7.3 What kinds of data are represented in them? (Source: graphic image supplied courtesy of Environmental Systems Research, Inc. and is used herein with permission)

☑ using maps interactively, to display the data, to generate new questions, to provide new views of the situation, and thus to produce better understanding.

Figure 7.4 Displaying data interactively (Source: graphic image supplied courtesy of Environmental Systems Research, Inc. and is used herein with permission)

Because of this,
geographers try
to figure out
why things are
where they are.

They analyze the
patterns of what
is where and
look for all the
why's.

Figure 7.5 Location and patterns (Source: graphic image supplied courtesy of Environmental Systems Research, Inc. and is used herein with permission)

All these events
and questions are
special because
they are spatial.

They happen in
some places and
not others.

We need to analyze
these situations,
focusing on their
spatial patterns.

Figure 7.6 Spatial analysis (Source: graphic image supplied courtesy of Environmental Systems Research, Inc. and is used herein with permission)

questions are special because they are spatial. They are open in some places and not others. We need to analyse these situations and focus in on their spatial patterns (ESRI ArcView Explorer, released 1992).

Geographical information systems are used by many people to try to solve spatial problems. This can be really helpful because problems are not as 'clear-cut' as they used to be or may not truly lie in one discipline or another. Deciding where the new landfill goes in or trying to plan for the location of a new business may require very different types of information upon which decisions will be made. GIS can help provide that information. It can provide maps which show how the data interacts. It can show patterns and relationships. GIS can even be used to query or ask questions of the data to show specific relationships or areas. GIS are a powerful tool for increasing decision-making abilities.

An example would be a community which is looking for a site on which to build a new landfill. Various factors would come into making this type of a decision. Where would the landfill be built that will be near a major road for transportation of the waste? What type of zoning is necessary for a landfill? What types of soils are needed to prevent leaching into groundwater? What type of buffers will be needed to keep the landfill away from nearby housing, businesses, creeks and forested areas? With a GIS package, each different type of information would comprise a data layer. By examining each data layer individually and together a possible site could be selected. The true power of the GIS comes into play when all of the data layers are queried. The software asks questions of the data and selects those parts that meet the questioning criteria. Those possible sites are then outlined on a map with their appropriate buffers and soil types listed. This is a much faster and more powerful way to find answers to complex questions such as landfill site locations.

Types of Geographical Information Systems available

There are a great variety of GIS tools available for people to use on a variety of computing platforms. There are also many types of mapping programs available which provide the user with some simple tools for dealing with maps and geographic information. Hardware for use with GIS has traditionally been in the mainframe, minicomputer and workstation level of computing. These machines have historically been the only machines with sufficient power and storage to deal with the vast amounts of information generated by the database and the mapping layers. Lately, Macintosh© and PC platforms have seen the availability of powerful GIS applications become available. Low-cost applications such as MacGIS©, MapInfo™, and Atlas Maps© for the Macintosh have become available, whilst PCs have seen several of the ones already mentioned as well as products like ArcView and PC-Arc/Info. These tools have a lot of power and provide full GIS capabilities on a PC platform. They are relatively affordable and have a modest to steep learning curve.

Student introduction to GIS systems

Students working with GIS should start with basic map and cartography fundamentals to understand the basics of map making and use.

They can be introduced to mapping via constructing maps of real-world places near them. Students probably have some sense of maps from geography or history classes already. They may need review over such construction practices in cartography as scale, legend, titles, borders and map themes. Some examples follow.

Map construction: initial steps and overview of the project

Directions

Students working in groups from two to six will use a surveyor's level to construct a simple boundary map for a local area of significance.

Reason

This project should be completed on an area that needs to be mapped. A product with a purpose should be generated. Possibly, the area is a playground and needs to be mapped for a safety review. It could be a marsh area that is being studied in another class, possibly a science or social studies class. The following equipment is needed:

- Surveyor's level
- Stadia rod
- Data sheet
- Compass

Procedure

MAPPING TABLE TECHNIQUE FOR CREATING MAPS

With only a table, paper, straight edge, measuring tape, and some marking stakes, using the mapping table strategy.

First you mark out a line (50 feet or 50 metres). Then on a piece of paper you mark the line (4 inches = 50 feet) – you choose the scale on the paper. You have someone walk the edge of what you want to map and then as they follow the contour, you draw a line to them from the endline – see dotted lines on figure 7.7.

You then change ends of your 50-foot or -metre line and draw lines to the points that were drawn to as you had someone walk to the points the first

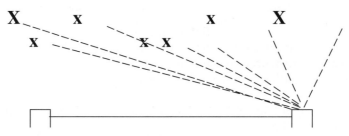

Figure 7.7 Using the mapping table strategy: part I

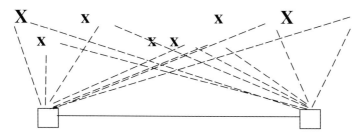

Figure 7.8 Using the mapping table strategy: part 2

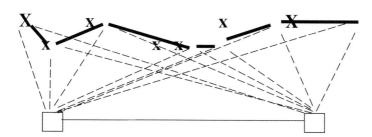

Figure 7.9 Using the mapping table strategy: part 3

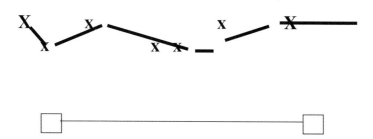

Figure 7.10 Using the mapping table strategy: part 4

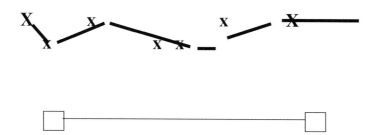

Figure 7.11 The finished map: the edge of Cow Creek at Longhorn Crossing

time. That means they have to walk by them again. You may want to mark them with stakes. You probably will want to use a straight edge to draw the lines to keep things straight (see figure 7.8).

Once drawn, then you can connect lines drawn between the X's and erase all of the lines drawn from the mapping blocks at the end of your 50-foot or -metre line (see figure 7.9).

This scale line then represents actual real-world distance and the contour of what you are drawing is to scale (see figure 7.10).

A finished map might look like figure 7.11.

MAPPING A SITE WITH A MEASURING TAPE

The strategy for measuring distance is as follows: in this type of mapping the area that you want to map is mapped using a measuring tape and a compass. A tape is used to measure the length of the site and left there as a reference. Then another tape is used to measure horizontal distances along the width at various distances across the site. When a scale is chosen the distances can be converted and measured on the paper. The top of the paper represents North and the map can be annotated or labelled (see figure 7.12). This technique is probably the easiest for younger students.

Mapping with Expert Maps

Expert Maps is a low-end mapping system that provides visual outline maps that can be annotated, printed, or saved as a graphic image for inclusion in a desktop publishing program. It is extremely easy to use and is menu driven. The cost for this program was less than US$20. The reason for starting students with this program is its simplicity. It is extremely easy to use and since it really only provides outline images of states and countries a lot of computing

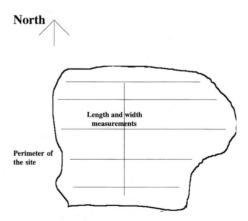

Figure 7.12 Mapping a site

power is not needed. This program does not produce legends nor is at any type of scale. What it does is provide students with the ability to start working with a program that will give them simple maps to accompany projects that they are doing. The program also starts students thinking about the visual image questions of what type of map do they want to show this information. Maps can be shaded, have text added, arrows can be drawn, and the map can be saved as a .pcx graphic file for inclusion into a paint or art program. Expert Maps is not true mapping, but it is a beginning and an initial way for students who may not know their way around a computer to start to generate ideas and produce a product.

Suggested activities

* Generate cover map images for report covers. The students generate a rough image showing the regional location on a map. This could also be a more local location too. The students then clip this image into a desktop publisher and then make a title page for their research report.
* Construct general maps showing location of a research project for others from a world-wide view to a more regional and local view. Students use Expert Maps to create the slideshow of maps that get the viewer closer and closer to the actual study or research site. Not much detail can be shown but shading and text can be placed on the map (see figure 7.13). This will lead the student into developing the intended message of the map and having some general skills in doing so.

Expert maps

These maps can be used to show general locations, give regional meaning to work being performed, and start to develop skills in map construction (see figure 7.14).

Cartographers should spend much time and detail on the design of their maps. Layout and design are important based upon what you want people to see and understand from your map. The use and types of colours are very important.

Mapping skills with AutoMap

AutoMap is a computerised road atlas with a lot of features that have some similarity to the layering of themes found in a GIS package. AutoMap is not a GIS package, but for the novice may be another instructional tool which can be used by having students of all ages working with maps, examining distances and routes, place data layers over the basic map (overlays), and be able to increase basic geographical knowledge about the United States and its features. Automap has an easy-to-learn interface and allows students to

Figure 7.13 Expert Map view of the Western United States (Source: graphic image supplied courtesy of Environmental Systems Research, Inc. and is used herein with permission)

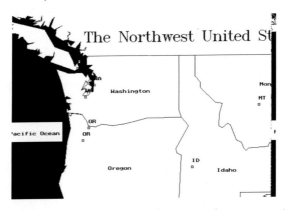

Figure 7.14 Northwest view with annotations (Source: graphic image supplied courtesy of Environmental Systems Research, Inc. and is used herein with permission)

start working with it right away. Its main feature is as a road atlas and that is what it does best. You can plan trips and AutoMap will calculate several routes to your destination as well as produce written instructions of directions to follow. It will produce a map with any of the many geographical overlays that it can generate. These are the overlays possible:

- State capitols
- Counties
- National and state parks
- Mountains
- Forests
- Lakes and rivers
- Indian reservations
- Military installations
- Game reserves
- Latitude and longitude grids

Data available

- State and city populations
- Land area
- Interstate miles
- Licensed drivers
- Registered automobiles/motorcycles
- Laws of the road
- Information hotlines for road conditions, tourist information, hotels and car rental

Details

- 359,220 miles of freeways, toll ways, state and county roads
- 51,921 cities

Possible activities

- Begin students by working through the road atlas portion of the program. Choose two cities and calculate the trip plan between the two. Usually, students will want to work from their home town to some other town. Have them calculate the top three or four routes to their destination, print out the map and the instructions for getting there. Several students may want to do this activity during holidays anyway as they plan trips to other places for vacation. Students can put together travel plans for themselves or possibly for the local travel agent. This is a possible partnership for students to work through these routes with travel agents.
- Students can start to develop maps which show geographic detail in their region. Using the zooms and moving the maps around students can centre the area they want to show on their map product. They can go through the overlay tools and add data to the map. They can also use the map as the basic layer and add different overlays to try and show different features. These could be printed individually, produced onto overlays to show a complete overlaid map or the map could be assembled in the computer and then printed out with all of the data.
- Students can work with students from other classes, especially geography classes, to help produce maps which fulfil individual or class needs for maps involving the US, states, highways, or any of the data layers (for an example, see figure 7.15).

ArcView: a GIS viewing program

ArcView for Windows is a windows-based program that has many of the functions of a full-featured GIS program costing many times the price. ArcView allows you to view and manipulate data which have been entered

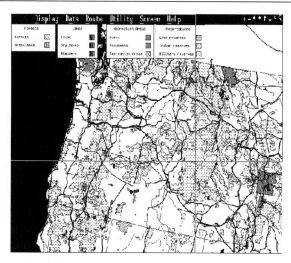

Figure 7.15 Automap view of the northwestern United States (Source: graphic image supplied courtesy of Environmental Systems Research, Inc. and is used herein with permission)

into ArcInfo, which is a full-powered GIS program. Arc/Info is the world leader in GIS mapping and so there is plenty of data available for ArcView. This program is Windows based and most of the hardware requirements are within reach for most schools. I have included the general information about GIS, ArcView, and the ArcView tutorial information which has been put together by ESRI (Environmental Systems Research Institute, creators of ArcInfo and ArcView). This material along with the tutorials which come with the ArcView package make learning the program fairly easy. There is a wide variety of data which is available for use with ArcView, including ESRI's ArcUSA, ArcWorld and ArcScene datasets.

Recommendations

ArcView is a highly recommended program for use with students. Because it runs under Microsoft Windows students will become comfortable with it quickly. Data can be obtained through ESRI's datasets (previously mentioned) or through agencies or state government. Anyone who is using Arc/Info GIS can set up data for you on ArcView.

ArcView requires a fairly powerful computer to run with any type of speed, and hardware requirements should be checked out closely. ESRI has a K–12 Education Program which is seeking to extend ArcView into the K–12 setting. ESRI has been giving the program away to qualifying schools. The required qualifications have been that a school has the necessary hardware to run the program and have a partner organisation which uses ArcView, PC-Arc/Info or Arc/Info. The organisation then acts as a mentor for the school. Schools interested in this program should contact ESRI at the address listed at the end of the chapter.

Projects and student interest

Students can learn to:

- Operate the Windows Operating System
- Use the tutorials in ArcView to learn basic GIS concepts
- Use the satellite scenes in ArcScene to learn about remote sensing and how to shift and process colour bands on the Thematic Mapper of the Landsat Satellite photos
- Obtain local digitised maps of the area and start to put maps together showing local points of interest
- Learn to use the Windows clipboard to process maps out into a paint program for adding information that is not in the database in ArcView
- Obtain road and natural resource information for generating layers
- Obtain available digital National Wetland Inventory Maps for ground truthing wetland types
- Use ArcView to generate maps showing species diversity and location for the GAP Analysis Project; these maps would be for local use and the data generated would be plugged into the full GIS at the state level and would be of great interest and usefulness at the local level
- Use ArcView to generate maps for local projects
- Use ArcView to determine important and rare habitats

ArcView Explorer

ArcView Explorer is a program which explains the function of ArcView and how GIS works. It explains how spatial analysis can be used to solve problems in the real world by looking at spatial data in new ways through a GIS system. ArcView Explorer is very friendly and does a great job as a tutorial for teaching basic GIS technology and concepts. Students should work their way through this software as a prelude for using ArcView for GIS manipulations. Figures 7.16 and 7.17 show examples of two screens from ArcView Explorer that illustrate how they explain GIS concepts.

ArcView map examples

Figure 7.18 shows ArcView's features in Windows. There is the table of contents that shows the themes, the tools window, and the map window. The data shown incorporate several different data sets. The Columbia River Estuary Habitat Zones, the major roads and minor roads are displayed. The software then has the capabilities to ask questions of this data. A possible

What sort of data is used in a GIS?

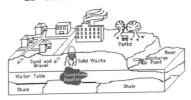

What kinds of data are represented here?

Figure 7.16 Sorts and types of data (Source: graphic image supplied courtesy of Environmental Systems Research, Inc. and is used herein with permission)

In a GIS, each category has its own "layer". Each layer can have many different elements. The total system stores lots and lots of data!

Figure 7.17 GIS layers (Source: graphic image supplied courtesy of Environmental Systems Research, Inc. and is used herein with permission)

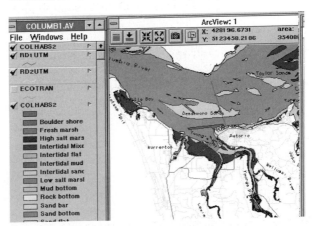

Figure 7.18 ArcView map example (Source: graphic image supplied courtesy of Environmental Systems Research, Inc. and is used herein with permission)

question could be: Where are all of the 'Rock Bottom' habitat zones in the Columbia River Estuary? Other questions could be asked just by having students zoom in from the initial map to more specific areas and features. This is what is shown in figures 7.19 and 7.20, which are just simple zooms from the initial map.

Zooming in on a map can bring out features hidden on the big map. Estuary habitat zones can be examined by students and compared with shore-line zoning for appropriate industrial uses in conjunction with land-use planning.

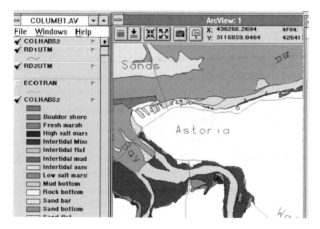

Figure 7.19 Zooming in on the Columbia River: example 1 (Source: graphic image supplied courtesy of Environmental Systems Research, Inc. and is used herein with permission)

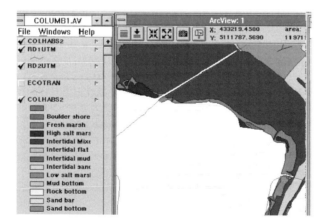

Figure 7.20 Zooming in on the Columbia River: example 2 (Source: graphic image supplied courtesy of Environmental Systems Research, Inc. and is used herein with permission)

ArcView materials

ESRI provides many materials for GIS in schools, for example:

- Geographic Information Systems for grades K–12: Using ArcView in the Schools
- Learning About Displays I – Windows
- Learning About Cartography, Part I – Colour
- ArcView 2.0 for Windows
- ArcView 2.1 for Windows and NUC

Ground truthing with GIS – an example: National Wetland Inventory maps

The National Wetland Inventory (NWI) maps are maps which show an inventory of our nations wetlands. Wetlands are extremely important for a variety of reasons and many of them have vanished owing to the pressures of development overtime. Students can play a large part in conserving and protecting wetlands by participating in the ground truthing of these NWI maps. The NWI map project was started by the US Fish and Wildlife Service in 1974. Their mission was to make an inventory and map the nation' s wetlands. The NWI maps are based upon aerial photography which was taken at 50,000 feet at 1 : 58,000 scale. They are overlaid over USGS topographic maps at 1 : 24,000 scale. Much of this information is used by various states for their own wetlands management programs. Students can obtain the NWI maps by ordering them with the forms which have been included in the appendix. Other sources from which students can obtain the NWI quads for their own area may include agencies or their state GIS Service Center. Some NWI maps may be in digital format but many are not and are only available in paper copy. Digital maps, if prepared for ArcView, can then be used directly by the software. Students can then use ArcView GIS and the digital data as the vehicle for producing their own zoomed versions of the maps and ground truth them.

Ground truthing means to verify the accuracy of the initial mapping effort by the US Fish and Wildlife Service. Students can use the quads to build maps and set up the project of verifying all of the identified wetland types on the quad and the local area. This information can then be sent on to US Fish and Wildlife Service as well as local and state planners who work with wetland issues.

Ground truthing these areas will allow students to work with the maps, develop new maps by manipulating the database, complete fieldwork and observations using their own and NWI maps, report results to concerned authorities who haven't been able to check the maps, and to use GIS as a tool to investigate and participate in the local society as concerned and informed

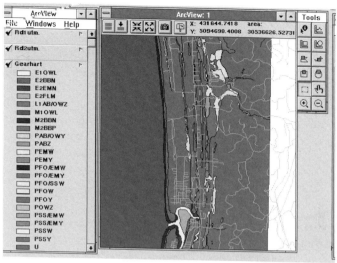

Figure 7.21   An example of the National Wetland Inventory (Source: graphic image supplied courtesy of Environmental Systems Research, Inc. and is used herein with permission)

citizens. At the local level, planners and natural resource managers are interested in updating and ground truthing NWI maps. The federal government is interested but will only accept information on NWI quads that are still in the draft review mode. Final map versions will not be updated in the near future.

Gap Analysis: GIS tools for tracking species location and diversity

Gap Analysis is a new project where the approach is to identify where the geographical areas (habitats) are that are critical to individual species or groups of species before their existence is threatened. This proactive approach uses students and citizens on the ground to map out where species are living so that the information gathered can be entered into a GIS system at the state level. One of the foremost efforts so far is the State of Washington with their GAP Analysis Project centred at the University of Washington. Oregon and many other states are following this lead.

In this project, students focus on certain types of organisms, birds, mammals, etc., and map out where they are found in their local communities. These census materials and maps are forwarded to the GAP Analysis people and entered into a GIS system. The habitat areas over the whole state are mapped and such things as wildlife corridors, flyways, etc. can now be discussed with much greater knowledge than before. Even if this was only taken on a local level, this type of project would be great to help in the local planning effort and eco-preservation efforts of local communities.

Oregon and other states are developing plans to begin their own Gap Analysis projects. Oregon's project will be co-ordinated through the Oregon Department of Fish and Wildlife (ODFW). Schools interested in participating in this project should contact ODFW's main office in Salem.

Students and teachers can participate in this project as well as groups of students. Schools from different areas can work together to put together maps. ArcView and the Windows Clipboard/Paint programs can be used to generate the initial wildlife maps. Working closely with the state GAP Analysis group could be a source for GIS maps and data to be used in ArcView for other GIS projects.

Project development: develop maps using local data

When working with students on GIS concepts, try to develop their mapping skills around local features. Developing a local project which integrates GIS as a tool is the best way to do this. Try to form a partnership with an organisation that is doing work locally and try to obtain local GIS data to work with. If these two things can be done with the same partnership then that would be an excellent situation. Students can do mapping and ground truthing work for projects like the National Wetlands Inventory maps as well as the Gap Analysis type of project. They can build upon these initial works with side investigations and use the GIS tools to look at the data in different ways. These maps then provide excellent support for these projects.

Much of the natural resource work which can be done and needs to be done is in the social science realm. It has to do with land-use planning, development and the direction and needs of society. Science and GIS then become the tools in which students can then explore these issues and participate as informed citizens in their local communities. The GIS potential for supporting integration between traditional subject areas is great and students should be able to support social studies departments with a wide variety of specialised maps.

ArcView GIS working with the Windows software allows for a lot of flexibility in map production. Data can be viewed and queried, maps can be cut to the Windows Clipboard and annotated via the Paintbrush or other art programs and incorporated into desktop publishing presentations.

There is a wide variety of data that can be used in ArcView. The sources for this can be found locally, through agencies, state agencies, universities, corporations and the federal government. Possible types of data are:

- Tiger files
- Road files
- Census data

- Estuary data
- Wetland inventories – NWI
- Voters registrations
- Forest types
- Zoning and land use
- Soils maps

Summary and conclusions

This chapter has explored some of the GIS training and use of the educational curriculum at the Coastal Studies and Technology Center, at the Seaside High School, Seaside, Oregon, USA. The curriculum is based around many projects and the work of a number of organisations and institutions. The objective has been to bring the technology into the 'hands' of school students, and more specifically to provide practical skills and understanding that enhance the educational experience with GIS. The practical project elements together with the links to scientists and agencies offers a unique combination of GIS technology and provides a sound basis upon which students will benefit both the scientific and social community.

Acknowledgements

Express Maps, Automap, ArcView, ArcWorld, ArcUSA, ArcScene, Arc/Info, Windows, ArcView Explorer, MacGIS, AtlasMaps and MapInfo are all trade marks of their respective companies and authors.

Many of the ideas shared in this chapter are collections of work which have been pioneered by others and whom I thank for the use of their materials.

- Environmental Systems Research Institute
- Washington State GAP Analysis Project
- Oregon Division of State Lands
- Columbia River Estuary Study Taskforce
- US Fish and Wildlife Service

Chapter 8

Expanding a corporate GIS into an authority's high schools

Stephen Gill and Peter Roberts

Since its initial conception in 1990 and implementation in 1993, Powys County Council has adopted a fully corporate approach to the implementation of GIS within the Authority. This has resulted in the development of a wholly integrated system across departments allowing extensive interchange of data (Gill, 1996). A logical extension of this approach was to implement this same system within the Authority's high schools.

This paper examines the approach adopted to meet this challenge. Particular attention is paid to the problems faced in terms of finance, training and data provision. The paper concludes by examining the achievements in the first year, the potential long-term benefits, further expansion opportunities within secondary schools and possible implementation in some primary schools.

Educational context

Changes to the National Curriculum in the UK, particularly in the geography syllabus, have resulted in greater emphasis on the integrated use of mapping and IT. In particular, GIS is recognised as a means of meeting these require- ments both in terms of using digital mapping and handling many types of data, including social and demographic tables frequently used by schools and students. The spatial analysis capabilities of GIS in two, three and four dimensions are appropriate and applicable to many specific requirements of the GCSE and A-level syllabuses. It had been a long-held goal of the Powys corporate GIS team to introduce GIS into schools together with a variety of digital data sets, including OS maps supplied under the local government service level agreement (SLA).

Powys County Council

Powys County Council (PCC) is a rural unitary authority occupying 2,000 square miles, approximately 25 per cent of the area of Wales. PCC services the needs of approximately 125,000 people with the largest individual settlement containing just 10,000 residents. Education needs are delivered via 109 primary schools and thirteen high schools dispersed throughout the county.

Over the past five years the Authority has been developing an ambitious, PC-based GIS implementation, networked across a very extensive Wide Area Network (WAN) into hundreds of workplaces and potentially available to thousands of staff (see figure 8.1) (Gill, 1996).

This corporate GIS initiative has been led by Steve Gill, Head of Strategic Planning and Information and GIS Manager, with a steering group drawn initially from staff based within Planning, Technical Services, Emergency Planning and IT departments. This approach has provided considerable improvements in management of resources, information efficiency and the effectiveness of service delivery for those departments. Geography teachers in the County's high schools were quick to realise the value of GIS, although to date the Education Department are relatively low-level users.

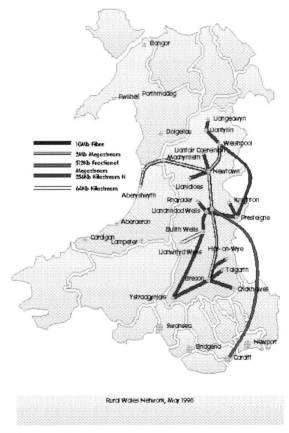

Figure 8.1 Rural Wales network

Digital mapping and spatial information on everyone's desktop

Powys' GIS goal was always a corporate one, involving as many services and users as possible right from the outset. The underlying reason for this was the recognition that the value of any data increases exponentially with the number of users that make their data available and the number of other services that can effectively utilise them. A corporate GIS provided the ideal platform for data integration, effective dissemination, avoidance of duplication and introduction of data standards.

Powys selected MapInfo™ GIS because it could be implemented on the existing Novell and Ethernet-based WAN utilising existing PCs and peripherals. Initially, a large new PC-based Fileserver was necessary to accommodate the map base and spatial data sets together with some more powerful PCs with large monitors. Some higher-quality large-format printers, digitising tablets and scanners were also essential.

Initially, GIS users located off the WAN had map bases installed on large hard disks; later map data and other essential data were supplied on CD-ROMs mastered in-house. This last development was to prove essential to the success of the schools implementation. As time passed, the WAN was extended and improved, and many more local servers were installed covering the very many distributed workplaces.

Developing the corporate profile

By the end of 1993 Powys had trained twelve GIS users, rising to twenty-four one year later. All short to medium targets were achieved in the first year. The original number of application areas had appeared ambitious, but actual GIS developments soon overtook the original plans. As staff see what GIS can do, they identify new applications for themselves.

In the third year the number of users doubled again, primarily due to the rapid take-up in Highways and Technical Services, which includes Consultancy, Streetworks, Streetlighting, Public Transport, Planning Liaison, Property, Estates and Valuation. At the same time, GIS developments became more flexible, e.g. running on laptops, and more sophisticated, e.g. utilising GPS (Global Positioning Systems).

In 1994 one of the three District Council Planning Departments, Radnorshire, joined the Powys GIS network and by April 1996, when Powys became a Unitary Authority, there were already about seventy-five MapInfo™ GIS users. Planning then became the main growth area and the total number of users quickly exceeded one hundred.

With so much of Powys on the MapInfo™ system, it is possible to share data horizontally across departments and between different offices in real time, ensuring that decisions are made based on the most up-to-date information at all times. In the Highways and Planning departments, for instance, almost everyone, including area offices located tens of miles apart, now uses MapInfo™ GIS over the WAN to access hundreds of datasets held centrally on the corporate GIS server.

This take-up from staff within former District functions has been carried through and during the 1997–8 financial year new implementations were established with the Land Charges, Housing, Environmental Health and the Treasurer's revenues section. By April 1998 PCC (not including the high schools) had over seventy-five GIS licences and more than 150 trained users.

The Ordnance Survey service level agreement

A major reason for the success of GIS implementations in UK local government has been the nationwide SLA with the OS. This gives all local authorities access to survey scale mapping (between 1 : 1,250 and 1 : 10,000 scale) of their area as a digital map base at concessionary rates.

The SLA also allows for data to be freely exchanged between local authorities, and PCC has taken advantage of this on many occasions, e.g. for cross-boundary projects and joint strategic working. The GIS team have also developed partnership arrangements with many other agencies active in the area.

Within the terms of the SLA, OS digital mapping can be made available to any school which is still managed by the Authority local management of schools (LMS) through the Education Department (see figure 8.2). This

Figure 8.2 Schools in Powys (Source: © Powys County Council)

initially only covered those products that are specifically within the SLA, e.g. Land-Line (vector) and 1 : 10,000 (b/w raster). Other OS products, such as Landranger Maps (1 : 50,000 colour raster) and Contours, may also be supplied, but only within the terms of the external purchase agreements for these products.

During the mid-1990s there was a trend towards encouraging individual schools to 'opt-out' of Local Authority control and become grant maintained schools (GMS), which are specifically excluded from the OS SLA. GM schools have established a separate umbrella agreement by which they can gain access to digital data (GIS in Schools, 1997; Ordnance Survey, 1997). Fortunately only one school in Powys (a primary) has opted for GMS status.

Under the OS SLA, LMS schools from other authorities can also use Powys OS data and vice versa. The former is quite a common occurrence in view of the number of Activity Centres in Powys and school field trips to the area, but these matters are not considered here.

Early GIS discussions with high schools

Most heads of geography in Powys' high schools had become aware of the Authority's GIS development over the period 1993–5. They were not only attracted by its handing of digital maps but were particularly impressed by its manipulation of demographic datasets like the Census and thematic mapping capabilities. This work was being developed by Mrs Di Greaves, a statistician in Stephen Gill's Strategic Planning and Information team who has become a leading GIS user, trainer and support officer within Powys CC.

During the mid-1990s, the stream of requests from pupils for maps and other GIS products grew from a trickle into a flood, including a trend towards incorporating GIS materials in school projects. The feeling was also growing that schools should really provide these services themselves to meet students' requests and also to create new course materials.

In mid-1995, a GIS demonstration to Powys Geography teachers on an INSET (in-service training) day confirmed everyone's belief that GIS had a very positive and creative future in high schools, but at that time there were no resources available to take the idea forward. These financial limitations were to prove the main barrier to implementation. At about that time, Peter Roberts joined Corporate GIS and his background in geography was to reinforce the intention to develop GIS in schools.

GIS and the National Curriculum

Schools within the UK are required to base their teaching structure within the framework of the National Curriculum. This ensures that core skills are developed by pupils in any particular topic, but leaves individual schools with a degree of flexibility as to how these requirements are actually delivered.

The curriculum in Wales contains a number of references to students being able to show their abilities to use a range of sources, including maps, and to depict or illustrate their ideas, analyses or conclusions using maps – see later references to Key Stage 3 covering the first three years (11- to 14-year-olds) at high school. In Northern Ireland, GIS has been explicitly written into the Curriculum at Key Stage 4 (14- to 16-year-olds, GCSE examinations) and a CD-ROM has been prepared by OSNI (Ordnance Survey of Northern Ireland) and launched to support this move (Galloway, 1997).

Other specific references to mapping in the syllabuses include mention of the study of uneven population distributions and other demographic characteristics. Some packages for schools such as Maps and People (Peebleshore, 1997) are specifically designed to meet these requirements using real Census data.

The geography syllabus obviously contains many additional references to students showing their understanding of map scales, orientation, symbology, relief and interpretation. Of particular interest to the GIS team is the specific encouragement of IT to aid the delivery of these requirements.

All of these concepts are more developed and detailed at A-level (16- to 18-year-olds). The use of mapping within submitted projects is encouraged by the Welsh examining boards at this level.

The initial opportunity

As Powys departments become GIS users they are encouraged to develop their own applications with the corporate GIS team providing advice and support. The Education Department became direct users in June 1998, but for several years before that used corporate GIS as consultants on one-off projects.

In 1996 the earlier discussions with geography teachers in the high schools took on a new lease of life with the involvement of Powys Training and Enterprise Council (TEC). Earlier ideas were worked up into a feasible and affordable proposition under the TEC's Science and Technology Programme.

The principal trigger for this was the special educational pack from MapInfo™ Corporation of sixteen licences for just £100 each, well below the normal recommended retail price (RRP) of £1095 and well within the envisaged budget. Aimed primarily at universities with large workstation rooms, it was confirmed that the offer could equally be applied to designated courses within schools that would lead to recognised qualifications. Powys were apparently the first Authority to explore and take advantage of this opportunity.

Project aims

The primary aims of this initiative were:

- to supply every high school in Powys with a copy of MapInfo™ GIS
- to provide members of staff with GIS facilities that would not consume excessive amounts of time, but which would help improve the quality of pupils' work and would help them to meet the IT demands of syllabuses more fully, particularly geography
- to train at least one teacher in each high school to use GIS
- to realise the full benefits of the OS SLA in the school environment
- to present and analyse statistical and spatial data within the GIS environment for use in the classroom and in project work

The project takes shape

Powys is a very large county and at that time (early 1996) was covered by two TEC Education Officers. They both came to see GIS working at the Royal Welsh Show at Builth Wells in July 1996 and were very pleased with the educational potential of what they saw. A further meeting shortly afterwards in County Hall saw the GIS in Schools Project receive financial backing from Powys TEC.

A meeting between the PCC GIS team, Powys TEC, the Education Department and an IT teacher representing the schools determined the detailed technical feasibility, data requirements and the proposed timetable. Teacher training was scheduled for the following two INSET days in early 1997.

Issues surrounding OS copyright were discussed and it was agreed that maps should not be printed off by unsupervised schoolchildren or used for non-educational purposes.

The use of the GIS licences was discussed and it was agreed that one would be located in each of the high schools and one in the special TEC/schools Partnership Centre in Welshpool, which supports primary schools in the northern area. Regrettably, there is no equivalent in the southern part of Powys.

This distribution clearly pointed to the use of each licence by a single user at any time, which might be a teacher (in class or doing research or preparation) or a single pupil or a group of pupils on a single workstation. This matter would be kept under review and reported back to future meetings.

Data delivery

The original intention for the project had been to distribute OS maps and other data using the PCC WAN to which all the schools are connected. However, this was beset by a number of technical problems, the most serious of which were the absence of connections to the network in classrooms where the data would actually be required and the lower data transfer capacities on the schools' own local area networks (LANs). This directly mirrored the earlier problems faced by dispersed PCC offices off the WAN and the same mechanism of data transfer, CD-ROMs, was adopted.

This option has a number of additional advantages: the CD-ROM offers data protection, being read only; it ensures that the software licence agreement is honoured because each school had just one copy of the OS map data; and it ensures a high degree of control over the OS copyright, because individual teachers from each school signed for each numbered CD-ROM and at the same time confirmed that they were aware of and accepted the OS copyright restrictions that had been explained to them.

Errors are known to occur in OS data and the co-operation of teachers (and pupils) in identifying and reporting these errors was sought.

Implementation

Powys TEC staff undertook the installation of a single copy of MapInfo™ GIS software in the thirteen high schools and the Partnership Centre, Welshpool, over a six-week period prior to Christmas 1996. At this stage the only data supplied were the demo data included on the MapInfo™ CD-ROM

At the same time a copy of a MapInfo™ training manual was left with the staff who were scheduled to attend the training sessions, giving them at least six weeks to familiarise themselves with the concepts of GIS in advance of a formal training session.

Training

Powys CC became self-sufficient in terms of GIS training provision as early as the end of 1993. All prospective users are required to attend one full day or two half day sessions with a corporately specified syllabus before they are allowed access to the corporate GIS file-server, which holds the Authority's map bases and spatial data. The training sessions have been developed around eight core modules, which combine to cover:

- the concepts of GIS
- the basic functionality of MapInfo™ GIS
- OS copyright restrictions
- data capture and creation
- best practice guidelines for GIS use

Users are then expected to further develop their skills by reference to training material provided with their copy of MapInfo™ and additional advanced modules supplied by the corporate team via an Intranet site. In preparation for the schools' training sessions, these modules were substantially revised to deal with a number of situations that would be unique to the high schools; in particular, the section on OS copyright was substantially expanded.

The first training session was undertaken by Peter Roberts and Stephen Gill at Llanfyllin High School on the January 6th, 1997. On this occasion training was supplied to ten members of staff from High Schools in the northern half of Powys, the northern TEC officer and a member of staff from the Welshpool Resource Centre. The teachers were primarily from geography departments with some IT colleagues, which was the suggested blend. They had a range of computer ability from very basic to advanced graphics skills, which did pose some problems as some struggled a little with fairly straight-forward tasks.

The second training session was undertaken by Peter Roberts and Chas Futcher, a planning technician, on February 24th, at Builth Wells High School. On this occasion a total of eleven teachers from the south of the county were involved. The training mirrored the first session in its content, but in addition Stephen Gill introduced Roger Geans of the OS Education Team, Southampton, who added a short presentation on the various support packages the OS have for schools.

Follow up

As part of the Powys GIS community, all trained members of teaching staff are now on the GIS mailing list and receive copies of the corporate GIS newsletter, *GISMAP News*, and any training briefings that are produced. They can also call on the corporate GIS team by phone, fax or e-mail to solve any MapInfo™ GIS problems that have arisen.

Since initial implementation there have been two follow-up cycles including site visits where appropriate. A third cycle was scheduled to commence in the summer 1998 term during which a re-supply was to be made of existing mapping along with a number of new datasets which were identified by the Authority.

Examples of GIS usage in high schools

When the project was initiated, it was expected that there would be some degree of time lag before the software was used seriously in the schools. In most cases, however, GIS take-up was fairly rapid. All of the schools involved recognised the potential of the software and most were producing mapping for classroom work straight away. In three schools GIS use was delayed by specific IT issues, one relating to staff, the others to hardware. In only one school have problems remained unresolved.

At Key Stage 3 (ages 11 to 14) the GIS software is being used to develop the concepts of scale and distances. It allows teachers to zoom in to a map, thereby changing the scale without pupils having the confusion of adjusting to different maps. In addition, standard printed output is being used in local area studies in most schools.

At GCSE and A-level some of the most able pupils are using GIS under supervision for individual or group projects. This enables students to spend more time on geographic analysis rather than actually drawing sketch maps.

A number of schools have taken the project further than was initially expected and they are already applying the software across the whole curriculum and at various age levels. At Gwernyfed High School two pupils have submitted GIS-derived mapping as part of their A-level dissertations, one looking at historic river-channel changes, the other the historic development of a rural village. At Newtown High School, GIS is being used in relation to (GNVQ) course-work.

GIS usage in primary schools

The Information and Corporate GIS Teams receive many mapping requests from primary schools; these cover a wide variety of uses from displays in school to local history projects, from orienteering events to environmental conservation, e.g. Agenda 21. At eight years old, Stephen's daughter is studying environmental designations such as sites of special scientific interest (SSSIs) and Local Plan maps in school. The school also supplies CD-ROM atlases to study UK and world geography, capitals, flags, etc.

There was no direct intention to supply GIS to primary schools at the outset, but there have been a number of spin-off benefits. Both the Welshpool Partnership Centre and Gwernyfed High School have used GIS to help local primary schools, e.g. by supplying them with village maps for use in locality awareness exercises at Key Stages 1 and 2 (ages 4 to 10).

Data issues

Since implementation the single largest problem to emerge has been data availability. At the outset a number of key datasets were identified and most of these were supplied on the initial CD-ROM. Subsequently teachers have requested a number of additional data sets for use within the GIS including:

- Contour information
- Census statistics
- Climatic/nature conservation information
- Global statistics e.g. for use with EU and developing world-case studies
- Historic mapping to illustrate settlement growth and river-channel change
- Integration of data supplied from other packages

Of these requirements contour data was available to schools from the summer of 1998 as a result of OS SLA changes and a global statistics dataset has been derived from the online version of the World Factbook published by the US Central Intelligence Agency (CIA). Options on the other requirements are still being explored.

At the same time other mechanisms to make more data available to schools are under consideration, including:

- providing schools with a discrete directory on the main GIS server
- providing a new dedicated schools server
- acquiring or developing off-the-shelf packages for specific educational purposes
- Maps and People or SCAMP (specialised Census information) (Peebleshore, 1997)

Each of these options have technical difficulties and security issues associated with them. In particular the security implications of providing access to our primary GIS server are likely to rule out this option.

In the medium term it is likely that CD-ROMs will remain the preferred delivery mechanism, especially with the improved data transfer rates of modern CD-ROM drives.

The implementation of a new schools' server to provide digital maps and spatial data as well as other datasets is a viable option, although substantial infrastructure improvements to LANs within schools would need to be made for this to be technically feasible.

The future for GIS in Powys schools

After one year, the Powys high school GIS project can be considered to be a success and has produced a number of significant benefits to teachers and to students, for teaching preparation, classroom work, community projects, individual project work and extra-mural activities.

Not all schools have progressed at the same rate, a fact that is not unexpected due to the wide range in IT ability displayed at the training sessions, and in one instance the software has not been used. The challenge now faced is to expand the scope of usage in the more advanced schools while helping the remainder to catch up.

In the advanced schools the availability of just one licence is proving to be a major constraint upon development – for example, preventing hands-on experience for groups or classes or preventing more than one teacher using GIS at any one time. In this instance the limited budgets available to individual schools are again likely to be a problem, although new advances in terms of map viewing technology are reducing the potential costs.

The challenge of the next phase will be to ensure that the learning experience is passed on to others and that mistakes are not duplicated. High schools in Powys are physically very distant from each other so communications and co-ordination are going to be important factors in building on the initial success. To this end it is likely that additional training will be required, probably through distance learning techniques. In addition, drawing together 'best GIS practice' as a 'training resource' will play an important role in expanding use for teachers.

The real potential for benefit lies in the pupils themselves, especially those working at A-level. While most of these are likely to go on to university, a number will end up in the local jobs market and the benefits to the Authority of having new staff who already have some GIS background cannot be underestimated when it comes to on-going GIS training costs. To this end the provision of a number of special half- or full-day sessions for A-level pupils is actively being considered at this point.

The logical extension of the project is into the Authority's junior schools. However, simplified or cut-down versions of MapInfo™ GIS might be more appropriate, together with some additional menus to aid navigation through the functions and to facilitate the selection of appropriate commands. Possible means of doing this at a fraction of the usual cost are now being actively explored.

Summary

Fully functional GIS software has been provided in every high school in Powys, wholly compatible with the PCC corporate GIS strategy. Potentially every high school can connect to the PCC WAN and become part of the PCC networked GIS community (Gill and Roberts, 1997a,1997b; Roberts, 1997).

In most high schools two teachers have been trained to use GIS, usually from the geography and IT departments. Each school was supplied with OS base mapping and other data on CD-ROM. The training included familiarisation with OS copyright issues. The feasibility of providing GIS in every high school has been proven.

The take-up of GIS was much faster than anticipated. Every high school, except one, has fulfilled the initial aims with most exceeding expectations, and the benefits have been amply demonstrated. GIS has been applied to very many aspects of the geography syllabus and to many other aspects of the curriculum. This provides a very encouraging foundation for further applications and full integration into the national curriculum.

Some technical and data issues remain to be resolved, but there is clearly enormous scope for further development of this educational opportunity. There is a huge willingness and commitment from all involved to undertake further development with GIS, but as usual in schools, funding could be a problem. Therefore, further development of GIS in schools must be extremely cost-effective.

Pupils on work experience have already worked alongside GIS users in the Authority and it will not be long before school leavers are listing GIS awareness and skills on their curriculum vitae when they apply for jobs with the Authority.

References

Galloway, B. (1997). 'Schools of Thought, Putting GIS on the Curriculum', *Mapping Awareness*, April, pp. 32–4.

Gill, S.G. (1996). 'Corporate GIS Implementation: Strategies, Staff and Stakeholders: A User-led Approach to Developing a Flexible Desk Top GIS to Integrate Data', *Geographic information – towards the millennium*, AGI, London, Paper 5.2.

Gill, S.G. and Roberts, P.D. (1997a). 'Powys High Schools Adopt MapInfo™ GIS', *GISMAP News*, Powys County Council, February, p. 1.

Gill, S.G. and Roberts, P.D. (1997b). 'GIS – The Next Generation: Every High School in Powys Joins the Corporate GIS Community', *Geographic Information – Exploiting the Benefits*, AGI, London, Paper 9.6

GIS in Schools (1997). *Geographic information*, May.

Ordnance Survey (1997). 'Digital Map Data – Special Price for Grant Maintained and Independent Schools', *Mapping News*, Vol. 11, p. 23.

Peebleshore (1997). 'Maps and People', a multi-media study pack including the SCAMP-2 CD-ROM, Peebleshore.

Roberts, P.D. (1997). 'Getting to Grips with GIS at School', *Local Government IT in Use*, Autumn 1997.

Further reading and information

Gill, S.G. (1997). 'Corporate GIS in Powys County Council', *Mapping Awareness*, May, pp. 22–5.

Gill, S.G. and Roberts, P.D. (1998). 'Use of GIS in Education – Powys County Council', *PITCOM Journal*, Parliamentary IT Committee, Spring.

Chapter 9

Another school of thought

Introducing GIS to a secondary school geography department

Stephen Walker

This is a personal view of the introduction of Geographical Information Systems (GIS) as it has applied to one school. The Holgate School is a co-educational age 11 to 18 comprehensive school with approximately 1,300 pupils on the roll. There are eight forms in each year group from year 7 to year 11, and a sixth form (years 12 and 13) of about 110 students. Most pupils live within the catchment area of the school, which covers the town of Hucknall, Nottinghamshire. Hucknall has a population of around 30,000 and is situated in the industrialised west of the county, close to the M1, and 12 kilometres north of the centre of the city of Nottingham (see figure 9.1). The school is funded through Nottinghamshire County Council, although it manages its own budget.

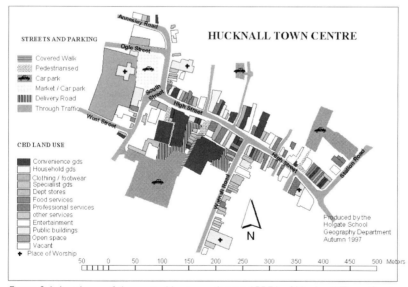

Figure 9.1 Land-use of the central business district (CBD) of Hucknall Town Centre

Geography is taught to all Key Stage 3 (KS3) pupils – in years seven, eight and nine. The subject is a popular option in Key Stage 4 (KS4) with around ninety of the 240 pupils in each year group opting to study GCSE geography (in years 10 and 11), and a further thirty opting for a non-examination geography-based course called 'local studies'. The Holgate Sixth Form is fairly small, and geography is taught to an A-level set of around eight students in each year group. Most geography in the school is taught by four staff, whilst six others teach a small component each.

The ESRI Schools' Prize: the GIS stimulus

Our introduction to GIS was neither planned nor expected. In 1995, the department won the ESRI Schools' Prize – a prize presented annually by the Association for Geographic Information (AGI) and sponsored by the Environmental Systems Research Institute (ESRI). The award was based on our submission for a project planned by and intended for Sixth Form students. The project was conceived as a fieldwork-based enquiry to study the changing use of green-belt land around the school.

Hucknall is situated on the northern fringe of the city of Nottingham. In recent years there has been increased loss of the green-belt, as development spreads from the city, and (particularly) towards the motorway junctions. Hucknall also lies within the concealed portion of the Nottinghamshire coalfield, and deep mining has been an important land-use in the past. Evidence of mining is to be found in many of the surrounding settlements and in the countryside. Recently, the search for greenfield industrial sites (partly to provide employment which will replace jobs lost from mining), the need for new homes, and the growing importance of opencast mining have all placed increasing pressure on the green-belt. Our original project submission was to collect and display data that would heighten awareness, and monitor the progress, of the changes to the green-belt.

A condition of being awarded the AGI/ESRI prize was that the project would be executed, and a display produced, for the AGI conference in November 1995. Within the space of a few weeks, a 'crash course' in GIS was undertaken and the deadline was achieved. Once this had happened the department had time to take stock. The school is extensively equipped with Acorn Archimedes microcomputers, and all pupils have information technology (IT) lessons using these machines. At the time we were awarded the prize there were only eight PCs in school – used in administration and the Careers library. IT was already part of geography courses, but it was based on the Archimedes machines, and did not include teaching about GIS in the accepted sense.

The AGI/ESRI prize consisted of a PC with 16mb RAM, a hard disc drive and a floppy drive. A copy of the ArcView GIS software was presented. The school already had a Canon 600 ink-jet printer, which is compatible with the machine, but is more used with other machines. Since the prize first arrived in school a CD-ROM drive has been added. At the time of the prize there was very little experience in the school with using a PC or with MS-Windows 3.1, and this was a slight problem in the early stages. However, the prize arrived at an opportune moment in the development of the department because we had recently agreed to introduce an enquiry-based approach to teaching geography.

Teaching geography

The aim of teaching geography can be summarised as making the pupils more aware of the world around them and to increase their skills by teaching them how to present, describe, compare and analyse data more effectively. The beauty of using GIS for this purpose is threefold:

- It can readily be used to demonstrate the inter-relationships between data sets
- It allows pupils to suggest, frame and investigate questions arising from the data
- Both of the above are carried out quickly

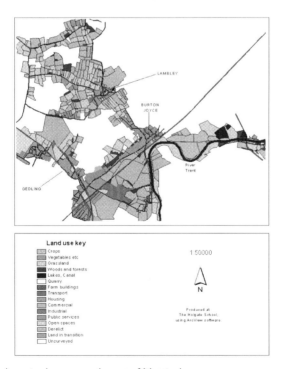

Figure 9.2 Land-use in the are north-east of Nottingham

In the author's opinion, GIS cannot be used to replace paper systems in their entirety. For example, pupils will still have to know how to shade a choropleth map, draw isopleths, or draw graphs by hand, but once they can achieve the basics and understand the use of each skill they can use GIS to automate the process rapidly.

The enquiry approach to teaching geography aims to encourage the pupils to describe and to try and explain phenomena from their own observation, rather than receiving most information from the teacher. They need to be taught to frame their enquiries. By using GIS the pupil can spend more time on critical thinking and less time on repetitive display and presentation. The correct use of GIS encourages more time to be spent in higher order skills like synthesis, questioning and evaluation and in this way learning should be enhanced. Like all other tools, GIS do not display or analyse data on their own. Rather GIS acts as the operator who must frame and carry out the enquiry. To pupils the initial appeal of using GIS is the novelty value, but later its use can promote deeper understanding.

Introducing GIS

The starting point for the introduction of GIS in the school should be an audit of existing schemes of work. In this case use of GIS had already started before the audit was undertaken. However, if starting from scratch the audit would have been the first stage. GIS is a tool whose use needs to be incorporated into current curricula. It can be used to support the existing learning and teaching objectives but should not form a separate entity. Like most departments the school had defined a 'progression of skills' – a list of the skills which it wished to introduce to pupils over the whole of KS3 and KS4 – in part based on the National Curriculum Orders. Many of these skills are progressive in the sense that mastery of one skill leads to other allied skills. The school began by reviewing the skills and techniques that it wanted to include in the geography courses, and then identified where GIS could be inserted. Consideration was given to where GIS datasets could be integrated into the 'places' and 'themes' components of the National Curriculum programme of study. However, as has already been suggested, the most suitable use of GIS is within geographical enquiries.

It is probably true to say that if the use of GIS is to take root successfully, then teachers must teach with GIS as well as about it. Presented with a single stand-alone PC, this would clearly limit the extent to which pupils could gain hands-on experience. No doubt the ideal situation is to have a classroom with several networked machines. Nevertheless it is felt that the single machine can serve a useful purpose in the classroom. The Holgate School concentrated, in the first instance, on three uses:

• For use by staff when producing teaching materials

- For demonstrating a range of data sets
- With the sixth form in their personal investigations

Having decided how and when to use GIS the next stage was to include a timetable of phased development into the Department Management Plan, and the inclusion of GIS within our teaching and learning objectives. At Holgate this occurred in March 1996.

In the first year after the award of the GIS time was spent getting to know the system. The staff had no previous experience of GIS and no competence with MS-Windows 3.1. However, newcomers should not be daunted by this; after all, presentation, description comparison and analysis of geographical data is an important part of any geography course from year seven to the sixth form. Staff and pupils are used to drawing maps; graphical skills are taught; and most pupils already have a working knowledge of word-processing and databases. Increasingly it is expected that school pupils will be familiar with PCs. The GIS is only special because it brings together these 'ingredients' within one system, and the basic medium is electronic. Within the limitations of the budget the Holgate School has embarked upon a programme of staff training.

The ArcView GIS experience

The school's experience is with ArcView GIS. This is an advanced GIS and is therefore a very complicated package. However, it is well supported by an on-screen 'help file' and the accompanying documentation. Gaining working knowledge is undoubtedly a slow process, particularly when one is in the situation of having no 'off-the-peg' teaching materials. Effective use of the GIS relies upon the existence and collation of worthwhile data. A collection of relevant project materials needs to be built up, and this takes time. Additionally, many of the techniques used by good geographers (like proportional circles, isopleth mapping or dot-density shading) are advanced techniques that are more difficult to achieve. Often it is a case of wanting to run before being able to walk! The difficulty is in gaining the necessary expertise whilst still carrying on with your existing teaching duties. Until GIS data and software are more widely available in schools this is not likely to change.

Availability of map data

An essential requirement of GIS is the existence of base maps. The system operates rather like an atlas with countless layers of overlaid material, so the base map is a vital starting point. Geography departments in state schools are fortunate that there is an agreement to supply some Ordnance Survey (OS) maps in digital form through the Local Education Authority (LEA). The Holgate School was fortunate that Nottinghamshire was very supportive in providing us with copies of digital map tiles. Without the licensing agreement, and the assistance of the LEA, this would be an expensive business. Even so, many LEAs are unprepared for schools who wish to request this data and you may have to be patient and persistent to gain the material that you want. Access to in-county map tiles allows the GIS to be used for local field work and will allow schools to create their own data set. Other data sets are available, from a variety of sources, but locating them may be time-consuming, and often they are expensive to buy. However, ArcView comes with a large data set that can be used to teach about global and North American applications.

Matching GIS to the school pupil's development stage

When considering the introduction of elements of GIS into the curriculum it is necessary to be careful to ensure that the tasks which are to be demonstrated or for the pupils to carry out, are suited to their intellectual development. There is a broad range of processes available for data handling. Some are too complicated for many pupils. We must clearly define the extent of the tasks and operations that we are to perform, if the pupils are to appreciate and understand what has been done. It is important to progressively introduce the skills and processes of the effective GIS operator. Most GIS can be 'customised' to remove unnecessary commands from the screen, and this will simplify use of the software.

Some example projects

Having considered the place of GIS within the curriculum, obtained the necessary hardware and software, and located suitable data sets, what happens next? Clearly, each department will proceed in a unique direction. However, it may be useful to consider some of the projects that have been undertaken at Holgate.

Land-use mapping

The original project, to map the use of land within the green belt around the town, was carried out hastily in the autumn of 1995, and then expanded in the following year, when the school volunteered to collect data for the Geographical Association (GA) Land-Use UK 1996 survey. Pupils and staff collaborated to collect data during the summer and autumn of 1996. Between us, through fieldwork, the school established a set of land-use data from a broad swathe of central Nottinghamshire, stretching from the River Erewash in the west and encircling the northern suburbs of the city, to Newark and the Trent valley in the east. Additional data came from other schools who had taken part in Land-Use UK. This information became the basis of a display mounted in November 1996 and was used for demonstrating data mani-pulation.

Campus maps

In year seven, the system has been used to create maps of the campus, and to model the effects of scale on the accuracy and extent of both maps and plans. GIS has been used to demonstrate basic calculations of distance and area. A useful exercise is to get the pupils to plot their journey from home to school, and then the system can tell them the distance. The emphasis at this stage is towards raising awareness and in having fun.

Urban land-use surveys

In year eight, the system has been used to collate and display the results of urban land-use surveys in the market town of Newark. The system has also been used to generate the base maps onto which data is recorded in the field. The software is used to illustrate the development and use of a database. Once again this is an example of the teacher using the stand-alone machine to demonstrate to the whole class. However, by processing the pupils' own data the exercise is made more personalised.

World map study

Parallel to our introduction of ArcView GIS, the staff also searched for oppor-tunities to develop the use of existing Archimedes-based software, one that would lay a foundation of the principles of GIS, and also allow the pupils hands-on experience in a familiar technological environment. It was obvious from the outset that at Holgate there would be a need to develop the use of the Archimedes hardware, because it is already installed! For some time the school has used 'World Map Study', a database/atlas package allowing for the display of global and continental patterns of data. A new teaching package based on this software was produced. It is an assessment task for year nine pupils, used in the 'global development and inequalities' unit, and carried out in collaboration with the IT department. This enshrines many of the principles of using GIS – for example, database handling, mapping at different scales and with varying keys, framing and displaying enquiries – though without the sophistication and complication of the PC package. The unit was successful, but there was a problem with providing differentiated support for all pupils. With further revision the unit will be used again.

Patterns of social deprivation

At the other end of the scale GIS has been introduced into the sixth form A-level course. It is used for teaching. An example is an investigation of patterns of social deprivation in Nottingham. Published data has been collated on social characteristics of wards from the 1991 census, health data (from the local Health Authority), data on the incidence and types of crime (from the Police Authority) and digital maps of the wards (from the county council). The students can frame and investigate research questions, and map their results.

Coastal geomorphology

In another example data has been collated about the Humber Estuary for a detailed case study within the coastal geomorphology unit.

'Personal investigations'

Elsewhere in the A-level course some students have used GIS as part of their 'personal investigation'. Within the A-level syllabus followed, up to 25 per cent of the marks are awarded for completion of a successful 'personal investigation'. The author has been very pleased at the extent to which relatively self-taught students (given some rudimentary training and the handbook) have managed to produce quite sophisticated results. One such example involved a student who mapped agricultural land-use in one parish, and compared this to published results from a similar survey carried out fifteen years ago. He created a data set using information interpolated from geology maps, topographic maps and interviews with farmers, and a database and spreadsheet records of his observations. The GIS was used to display the results. The project questioned the extent to which various factors appeared to affect choice of crops. This was a successful investigation and will be used as a model for future development of GIS in the sixth form.

Groups within the local community

Another approach has been to establish projects with groups within the local community. The school was approached, through a contact within the local police force, to establish a joint project to provide assistance to neighbourhood watch groups in plotting and displaying the incidence of crime. They were kind enough to allocate some money to upgrade the prize PC by the purchase of a CD-ROM drive. Permission was obtained from the local neighbourhood watch co-ordinators to collect and use their data. During 1996/7 one of our sixth form students based his 'personal investigation' on the collection, display and analysis of the crime data, and comparison with other potential factors. This has provided a useful community link for the school and an interesting basis for a project. Similarly the school is discussing with one of our local rural parishes how it might help them to produce a 'parish map' and with another how it may assist them to produce a 'parish footpath guide'.

Other departmental links

Within school, links with other departments have been explored. The technology department are interested in GIS as an example of a practical application of IT. The history department are interested in recording and displaying information relevant to local fieldwork (for example, former uses of buildings, and nineteenth-century census results). With the biologists, consideration has been given to a joint ecological study using data from the local forestry plantation (and perhaps using Global Positioning Systems to record the positions of individual specimen trees).

Developments

Involvement in Land-Use UK 1996 has already been mentioned earlier. Collaboration with ESRI allowed us to have a commemorative A0 poster of the results printed for display. As a sequel to this it was agreed with the AGI to host an event to mark Geography Action Week in November 1996. The event (which we called 'Geography in Action') saw several organisations who used or promoted the use of GIS collaborate in a day-long exhibition/demonstration for sixth-form students, teachers and interested members of the public from Nottinghamshire. In the first instance, the aim was to raise awareness of the nature, scope and potential of GIS in a range of applications applicable to schools. A secondary aim was to inform (and enthuse) teachers from local schools about the potential of GIS as a learning tool, and to raise support for co-operative ventures in the future. Thirdly, the school wanted to provide students and teachers with hands-on GIS experience, which would highlight the potential of careers using geographical skills and knowledge. A link was forged with the local branch of the Geographical Association (GA), who sponsored an evening session aimed at the general public.

It was a prime opportunity to see powerful displays from a software manufacturer (ESRI), GIS users (Diamond Cable Company, British Gas/Transco) and an academic department (the Department of Geomatics at the University of Newcastle-upon-Tyne). Although the event was only on a small scale, it was also useful, and the school successfully publicised and demonstrated the value of GIS within commercial organisations, and raised the awareness of the potential of GIS in schools. It introduced staff and sixth formers to the potential of GIS for their work and the extent to which it is used commercially – and hence potential career applications. The event also served to identify a small group of like-minded teachers and will form the basis for collaboration in the future. The author commends this exercise as a useful approach to building valuable contacts which will further the development of GIS.

To be successful, schools will have to explore links with local firms and businesses and try to seek assistance and expertise in their local community. At the Holgate School there are existing school-business links and these have served the school well. Several of the firms, who provide work experience placements for our pupils or who visit the school for 'insight into industry' events, have now been approached. There is interest within the business community to help schools in this way. The development of GIS is potentially very expensive and time consuming. It is worth exploring whether local firms can assist financially, or with data or expertise. It is also worth developing consortia with other schools to produce materials, if you can identify like-minded teachers. This could be a loose collaboration or establishment of a more organised user-group.

The future

So to the future. The Holgate School hopes to continue the development of GIS. Already funding is being sought to try to develop more GIS material in the 11-to-16 curriculum. Community liaison projects, to raise the profile of digital mapping and its importance, and partnership arrangements with other schools, with a view to developing a wider range of teaching strategies and learning material, are now being sought. Development is necessarily slow but the author is confident that progress is being made.

Further reading

Barnett M. and Milton, M. (1995). 'Satellite Images and IT Capability', *Teaching Geography*, Vol. 20(3), pp. 142–3.

Davidson, J. (1996). 'IT and the Geography Department', in *The Geography Teachers Handbook* (Eds P. Fox and P. Bailey) Geographical Association, Sheffield.

ESRI (1995a). *Explore Your World with GIS*, Environmental Systems Research Institute, Redlands, CA.

ESRI (1995b). *GIS in K-12 Education*, Environmental Systems Research Institute, Redlands, CA.

ESRI (1995c). *Exploring Common Ground: The Educational Promise of GIS*, Environmental Systems Research Institute, Redlands, CA.

ESRI (1996). *GIS in Schools: Infrastructure, Methodology and Role*, Environmental Systems Research Institute, Redlands, CA.

Freeman, D., Green, D.R. and Hassell, D. (1994). 'A Guide to GIS', *Teaching Geography*, Vol. 19(1), pp. 36–7.

Hassell, D. (1996). 'Using IT in Coursework', *Teaching Geography*, Vol. 21(2), pp. 77–80.

Holmes, D. (1996). 'Visualisation Software: A New Aid to Learning', *Teaching Geography*, Vol. 21(3), pp. 132–4.

Sheppard, S. (1995). 'Implementing IT in Schools', *Teaching Geography*, Vol. 20(1), pp. 17–19.

Warner H. and Hassell D. (1995). *Using IT to Enhance Geography*, NCET/GA, Coventry/Sheffield.

Warner H. (1994). 'CD-ROM Technology in Geography: Potential and Issues', *Teaching Geography*, Vol. 19(4), pp. 184–5.

Chapter 10

GIS and remote sensing in primary and secondary education

Rationale, strategies and didactics

Ove Biilmann

The coherent development of software, databases, primary and secondary school classroom practice, and teacher in-service education, in Denmark was conducted by a team of five to six persons over almost a decade (e.g. Biilmann et al., 1994; Biilmann, 1995a; Hubbe, 1988; Larsen, 1988). Studies of the development process and outcome were also carried out. The latter was often conducted in restricted study settings and sometimes dealt with only small teacher or student populations. Upon reflection, a great deal more could have been achieved and many important questions were unfortunately left more or less unanswered. The following considerations nevertheless reflect some achievements and indicate some of the constraints and questions inherent in this kind of development as well as in implementation in broader educational settings. This includes areas where the use of GIS and remote sensing has been successful, as well as areas where GIS/remote sensing may be of minor, if any, use. Elementary delimitations of that sort are prerequisites to further assessments or the development of perspectives and content areas where GIS/remote sensing are a means to achieve general educational aims and goals. Also, educational contexts, as well as frameworks for educational planning and practices, are discussed for heuristic as well as practical purposes. Consideration of mapping and map reading skills inherent in the interpretation and understanding of remotely sensed information and representations thereof is an important part of GIS/remote sensing didactics. Representations, as well as educational tasks of different complexity, are outlined. The emphasis is on rationale and outlines of development are faced with knowledge elements and skills ranging from elementary to complex and even advanced. At this stage it is important to refer mostly to the stepwise development of skills and knowledge or to content areas, which are easily recognised by all geography educators.

GIS enables students to navigate on maps and to manage different sorts of basic spatial analysis and representation. This is normally a mere reinvention of atlas use or map reading, supported by information technology. These sorts of GIS may add major flexibility, extra information, updating and the like to traditional map and atlas use. However, the basic educational challenges are more or less the same as before. This probably applies to map reading and mapping skills as well as to students' handling of the different scale and coverage of maps. It must be stressed that the capability of human beings in these areas are prerequisites in many daily life situations as well as to a variety of geographical analytical tasks and understanding of their outcomes. In any case, didactical analysis and development may also deal with questions of e.g. functionality and readability as well as of information access and accessibility. However, if the latter had been seriously considered, GIS/remote sensing might have been less popular by five years ago. The figures below support didactical assessment of these and other issues.

Some experienced limitations, which might limit or qualify the educational use of GIS and remote sensing, must be touched upon next. The didactics of GIS, as well as educational practice and lesson planning, may consider and cater for that sort of experience, however pedestrian they may seem. As an example, atlases are even more indispensable than usual, when GIS are used in education. This applies to primary and secondary education as well as to teacher in-service training. Not surprisingly, the PC-based GIS representations cannot yet compete with the accuracy, the organisational strength and the rich fund of information provided by a good atlas. On the contrary, the representations, comparisons and analysis provided by the GIS highlighted the atlas qualities as a handbook, the geographical organisation, source of viewpoints and systematics. Thus, the atlas and GIS are complementary sources of information through their mutual support of students' exercises and their enrichment of the educational content. Similar comparisons between GIS/remote sensing and other educational sources or media, e.g. textbooks, handbooks, topographical maps or statistical sources, are inherent in didactically-based work such as software design and specification, educational development, curriculum development, lesson planning, and teaching practice.

Remote sensing and beyond

A variety of maps (e.g. screen dumps), diagrams etc. exemplify some educational possibilities attached to GIS. This has been illustrated through various examples of 'good educational practice'. The current educational GIS approach is based upon software functionality, information access, 'good practice', and both commercial and political support. The didactical base and content development are often intuitive and exploratory. This has been confirmed by teacher interviews and other modes of evaluation. Given this state of affairs, other perspectives on geography and cross-curricular education, as well as of supplementary software dimensions, were gradually taken into account. This applies to the inclusion of elementary image processing and sorts of 'workbook' or hypertext features in the GIS software and to attempts to employ assessments in different educational contexts. The latter, amongst other things, applies to an endeavour towards coherent remote sensing and field study developments in a primarily geographical education framework as inspired by the work at the Aspen Global Change Institute (Biilmann, 1995c, pp. 7–11 and 31–6; AGCI, 1992). These efforts towards a broader methodology and content approach were implemented parallel to the latest stage of the IMPACT II programme of the European Commission, DG-XIII activities. It was more or less a case of revisiting classical geographical practices by virtue of the coherent use of old and new methodologies and information sources rather than a strictly new approach.

Figure 10.1 illustrates remote sensing as a representation of both place and landscape. The scale is diminishing and the size of regions is growing from the left-hand to the right-hand side of the figure (i.e. lower resolution or larger area per pixel). This highlights the continuum from place through regions of growing size toward continental or world coverage as well as the inherent diminishing ability to represent internal regional variation. Geographically interrelated dimensions such as source, scale, and size of region are inherent in remote sensing as well as in survey and field study. This is, for example, crucial in ground truthing, an aspect of professional remote sensing, as well as an educational strategy. Remote sensing and GIS provide both classroom- and field-study-based geography with new potential by including a powerful approach for coherent study embracing a variety of dimensions: probably a decisive strengthening of the geography education tradition as well as a potential for widespread support for geographically-organised analysis and representation in virtually all sorts of the world or 'reality' studies.

Didactic tasks such as choice, organisation and presentation of content are guided by educational aims and goals and conducted according to domain and educational theory and knowledge. Phenomena such as, georeferenced infor-mation, GIS software and ground truthing have been sketched in an attempt

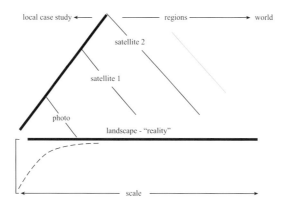

Figure 10.1 Illustration of remote sensing as a representation of both place and landscape

to clarify and facilitate didactic choices and specifications. Important aspects from remote sensing and GIS didactics is mentioned in figure 10.2.

In Figure 10.2, links of importance for design and development are shown on the left-hand side of the figure. Learning or teaching and development and their evaluation are represented on the right side. A specification of links and boxes is inherent in a continued iterative analysis. That leads to a differentiation of the boxes into a complex network with many two-way or feedback connections. A specification of domain and goals must be worked out in every case. Ground truthing and national land use morphology should be dealt with as exemplary strategies, where remote sensing and field observation are integrated into school subjects or projects. Developments and studies thereof are prerequisite rationale, content and study areas. They facilitate comparisons between students – studies of environment, landscape or region and of their studies of remotely sensed images of identical or similar landscapes and environments.

The addition of various features to GIS may also widen the educational potential. The Windows' features strengthen students – reading and responding to exercises by virtue of support of their writing process.[1] Figure 10.2 also illustrates different ways to introduce and utilise GIS and remote sensing in geographical, geopolitical and environmental education. The work of Poul

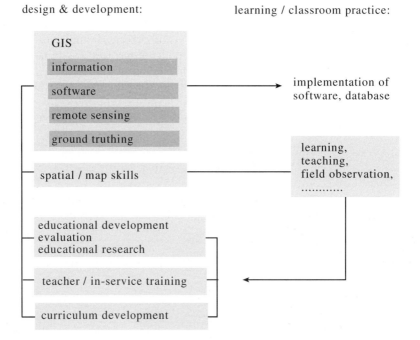

Figure 10.2 Important aspects from remote sensing and GIS didactics

Brondum, Keld Juhl Laarsen and others are based on a strategy for coherent development and classroom implementation of GIS, image processing, database access and an 'electronic workbook' (Biilmann et al., 1994). It was based on earlier remote sensing, field work and educational development projects. The uses outlined so far are applicable to classroom practice in Danish and other European contexts.

Rationale

We hypothesise that the issues above are inherent in many studies of information technology in geographical education. These issues may well be of some interest to general didactics of geography as well as to attempts to express and employ the interrelated dimensions of the geographical tradition in didactical practices. More than five years' work with these didactical issues provoked some careful rethinking. We therefore wish to share some of our earlier and recent considerations of rationale as a didactical issue with the reader.

We consider it certain that remote sensing, imagery and GIS represent an important if not decisive potential for geographical learning and core content. Free and easy access to relevant information must be added to the prerequisites already mentioned. We maintain that educational remote sensing and GIS share content structures, strengths and constraints embedded in geography and environmental education traditions.

We started to impose satellite imagery and GIS on classroom practice as examples of area or region representations primarily contributing to description. We hoped that more varied and flexible representations, e.g. better description – might work as a vehicle for an elementary analysis. Thus, we were looking for both student and teacher responses to GIS use. In fact GIS often was – and probably is – used as a magic, but black, box. It is sometimes expected to produce not only representations but also answers, or even solutions. This experience conforms quite well with GIS use conducted by some decision makers or their professional advisers.

Consequently, an IMPACT II definition project, didactical planning, considered GIS a compensation for lacking knowledge and skills rather than a tool for enquiry. That turned out to be convenient if not necessary (not least for raising interest, support and funding) but in some respects also short-sighted. Nevertheless, GIS was to a certain degree educationally legitimised along these lines. This aspect of the rationale was both confirmed and rejected one to two years later, as 11- to 12-year-old pupils in Odense explored a small Africa database by virtue of the 'black box' in a primarily enquiry-based education (Biilmann, 1995a, pp. 55–76).

Further didactical efforts and education implementation based on 1993–4 achievements may transcend this commonplace, but now established, rationale and contribute to a rationale for genuine educational remote sensing and GIS. The same applies to didactically-based information and media development (Biilmann, 1995a, p. 163) and further research efforts. It is stressed that the fund of psychological/geographical knowledge on e.g. map and mapping skills, survey and navigation has an important part to play in this area (Biilmann, 1995a, pp. 71–3; Blades and Spencer, 1994; Blaut, 1991). A look backwards provides further illustration of a developing rationale.

The introductory IMPACT II project educational GIS rationale (Biilmann, 1994b, pp. 218–21) suggests two areas, where geographical information and analysis are important: first, contribution to and support of geographical, environmental, geopolitical etc. knowledge, understanding and responsibility; second, a compensation for citizens', students', decision makers' and library users' lack of knowledge and skills.

This rationale and the derived educational software and database will probably lead to continued and new software and database development. Continuing didactical and educational developments are started. As the first rationale element is commonplace in geography, the following paragraphs emphasise the second. It needs to be specified, further developed and supplemented or exchanged by other rationale elements. Explorative ideas of that kind are outlined elsewhere (e.g. Biilmann, 1995a, pp. 173–4).

Continued efforts may include didactical review and discussion of GIS impact on geography raised by Curry (1994) and about the nature and function of images in geography (Phillips, 1993). Classroom experiences (e.g. Biilmann, 1995b, pp. 42–4, 47 and 77–9) suggest that even basic notions such as spatial similarity and familiarity need further exploration (Gale et al., 1990). Findings from this study are relevant for field work or ground truthing as well as for associated GIS use. Findings of connections between name and spatial location or between visual familiarity and location with similar regularities on a larger regional or even continental scale contribute to the didactical base for remote sensing and GIS. The shared educational qualities of remote sensing and ground truthing (AGCI, 1992) and key content areas of geography and environmental or local studies are crucial for continued educational development as well as for the further development of classroom geography. The same applies to learning where students integrate 'real life' or 'world' experiences in geographical and environmental learning – important learning stages aiming at the students' acquisition of balanced and realistic images of the landscape, region and world.

Appendix 10.1

Didactics

Didactics is concerned with educational goals, content, media etc. Didactical theory and practice attempts to grasp, contribute to and influence educational scenarios and contexts. It is based upon tradition and practice rather than on a coherent body of knowledge. Ontological or political assumptions are inherent in didactical theories and frameworks. Didactical practice is often normative and depends on wishes and possibilities rather than on confirmed knowledge or systematically collected experience. Didactical efforts always embrace compromises and may sometimes contradict experiences or traditional wisdom.

One or two building blocks of current didactics[2] of remote sensing and GIS have been outlined. This didactical area shares important dimensions with the geographical domain by virtue of the spatial analytical, psychological, man-land etc. perspectives. The directions mentioned above echo a growing psychological and educational awareness of knowledge and skills as domain specific rather than as general educational features (Johansson et al., 1985). Educational software and information access, as well as adapted textbooks and teacher support, should be combined or produced by virtue of closely connected developments based on a thorough didactical framework. This statement might overemphasise the differences between thoroughly didactically-based[3] strategies and approaches primarily based on application – e.g. modification – of professional GIS packages to classroom use. The latter approach may be to a larger degree feasible, successful and common as time goes by – for example as experiences and findings from didactically sound developments and evaluations are communicated and used. Nevertheless, it is still recommended to aim at coherent and mutual supporting software development, traditional publishing and information provision.

Acknowledgements

I am indebted to Poul Brondum and Keld Juhl Larsen for their considerable help in illustrating and reviewing this paper as well as for their permission to use some of their earlier work.[4]

Notes

1 As outlined in Biilmann et al., 1994b, pp. 343–50.
2 Relevant aspects of the concept of didactics are outlined in Biilmann, 1995a, pp. 163–4.
3 The notion of didactically determined information provision or software development may easily remain a theoretical rather than real phenomenon. This state of affairs is caused by economic, institutional and curricular constraints.
4 Our attempts to continue this development and research may fade, while the management of the Danish School of Educational Studies have handed the IMPACT II project finalisation over to a telecommunication company. The figures presented in this chapter are either derived from software versions developed when we were still responsible for didactics, educational development, systems description, dataset edition and processing in the project or represent outcome of later work in other contexts.

References

AGCI (1992). *Ground Truth Studies Teacher Handbook*, Aspen Global Change Institute, Aspen, CO, 143p.

Biilmann, O. (1994a). *Enviducation Educational Development and Evaluation Report 1993 No. 1*, The Royal Danish School of Educational Studies, Copenhagen.

Biilmann, O. (1994b). 'GRID-DK Smaskrifter 1', *Didaktiske noter*, Danmarks Laererhojskole, Copenhagen.

Biilmann, O. (1995a). 'GIS in General Education. A Didactic Framework', in *Proceedings* ScanGIS'95 (Ed. I.T. Bjorke), pp. 162–76, The Norwegian Institute of Technology, University of Trondheim, Trondheim.

Biilmann, O. (Ed.) (1995b). *Enviducation Educational Development Report 1994 No. 2* The Royal Danish School of Educational Studies, Copenhagen.

Biilmann, O. (1995c). 'GRID-DK Smaskrifter 3', *Didaktiske bidrag*, Danmarks Laererhojskole, Copenhagen.

Biilmann, O., Brondum, P., Hubbe, T. and Larsen, K.J. (1994). 'Information Access and Use: User Friendly GIS in Environmental Education', in *Europe and the World in Geography Education* (Ed. H. Haubrich) IGU Commission Education, Geographiedidaktische Forschungen, Vol. 25, pp. 339–51.

Blades, M. and Spencer, C. (1994). 'The Development of Children's Ability to Use Spatial Representations', *Advances in Child Development and Behaviour*, Vol. 25, pp. 157–99.

Blaut, J.M. (1991). 'Natural Mapping', *Trans. Inst.Geogr*, N.S. Vol. 16, pp. 55–74.

Curry, R.C. (1994). 'Image, Practice and the Hidden Impacts of Geographical Information Systems', *Progress in Human Geography*, Vol. 18(4), pp. 441–59.

Gale, N., Golledge, R.G., Halperin, W.C. and Couclelis, H. (1990). 'Exploring Spatial Familiarity', *Professional Geographer*, Vol. 42(3), pp. 299–313.

Hubbe, T. (1988). 'Local studies and the Use of General Application Programs in Ordinary Classroom Geography', in *Developing Skills in Geographical Education* (Eds R. Gerber and J. Lidstone), pp. 142–6, Jacaranda Press, Brisbane.

Johansson, B., Marton, F. and Svensson, L. (1985). 'An Approach to Describe Learning as Change Between Quantitatively Different Conceptions', in *Cognitive Structure and Conceptual Change* (Eds L.H.T. West and A.L. Pines), pp. 233–57, Academic Press, Orlando, FL.

Larsen, K.J. (1988). 'The Use of Computers in Local Environmental Studies in the Lower Secondary School, in *Developing Skills in Geographical Education*, (Eds R. Gerber and J. Lidstone) pp. 135–41, Jacaranda Press, Brisbane.

Phillips, R.S. (1993). 'The Language of Images in Geography', *Progress in Human Geography*, Vol. 17(2), pp. 180–4.

Further reading and information

National Science Foundation (1994). Global Learning and Observations to Benefit the Environment, *Announcement* NSF 94-152, NSF, Arlington.

Rudd, M. (1993). 'IT and Special Needs', *Teaching Geography*, 1993(2), pp. 84–5.

Sorensen, E.M. (1995). 'Geografiske Informationssystemer', in GRID-DK Smaskrifter 3, *Didaktiske bidrag* (Ed. O. Biilmann), pp. 13–29, Danmarks Laererhojskole, Copenhagen.

Chapter 11

The role of ArcView and Map Explorer in a corporate GIS environment

Stewart McCall

The use and application of GIS technology within South Ayrshire Council is still at an early stage. Indeed it was only with the demise of Strathclyde Regional Council in 1996 that the Council acquired any sort of GIS technology, albeit a DOS-based package called FASTMap. It was a further year after this that the Council began its use of ArcView, initially 'through the back door' as part of the Norsk Property Database System (UNI-Form).

The main drive behind the implementation of GIS in the Council was the new capabilities it could offer the organisation including improved efficiency in dealing with and using land-related data. It was also hoped that GIS would improve efficiency, lead to better decisions and improve integration between services.

Unfortunately this early optimism about the potential of GIS was soon to evaporate and the Council's initial steps were typified by a lack of understanding of the unique nature of GIS itself. The Council's failure to realise that GIS was not just another software product led to each department being left to do what they wanted and resulted in an unstructured, incoherent development of the system with the GIS failing to live up to its potential or hype.

A corporate approach

Although there are many documented examples of successful departmental-based GIS it was felt that the departments in South Ayrshire did not have the necessary skills or expertise to make such an approach work. Based on this, a decision was made to adopt a corporate approach to developing the GIS with the recruitment of a core team of GIS staff to develop a GIS Strategy for the Council and support GIS developments Council-wide. A corporate approach to GIS also fitted better with the management strategy of the Council, including the reduction of departmentalism through the sharing of resources, costs and information.

While the main focus of the resultant GIS Strategy was the user it was realised that the technology and how well it is aligned with the Council's data and organisational requirements was an important determining factor in the approach to GIS implementation. The Best Value initiative, to which all Local Government is subjected, also made the Council consider all the running costs of a GIS, particularly the cost of data, not just the on-paper costs of software and hardware. It decided that reducing the high per-seat cost of GIS was a key Best Value goal and this could only happen if GIS was delivered to the desks of as many staff as possible. The Council also realised that different staff would place different demands on a system – some staff wanted to use GIS for complex processing, data capture or analysis, while others only wanted to view and query spatial information. This and the Council's long-term goal for spatial information to be made available to the wider community led to two different software solutions being proposed.

ArcView and Map Explorer in South Ayrshire

The two products the Council decided to use were ArcView and Map Explorer. Using two products not one did not compromise the ideals of a single GIS solution as both products use the same data formats, can access the same data, and work in similar ways. The main difference between the two products is in their functionality: one is a complete GIS solution, while the other is a spatial data browser with read-only access to the data.

The core GIS solution for the Council is ArcView. ArcView is a fully functional desktop GIS and is used as the main GIS of choice for a small group of staff who are required to perform data management and complex data analysis tasks. This group of users could be referred to as 'power users' and for them the integral nature of GIS to their work requires a solution that is functionally rich, robust, and tailored to their needs. This group of users, although small in number, is spread across all departments, managing and maintaining the information collected by each service.

In contrast to the small number of users supported by ArcView, the GIS solution that has enabled more staff to benefit from the Council's spatial information database is Map Explorer. Map Explorer is an easy-to-use yet feature-rich product that allows anyone to explore, analyse and visualise geographic data in their work. Within the Council, Map Explorer is used by staff who are either currently unable to justify the expenditure on ArcView or who initially do not need the higher level of functionality ArcView provides. In many departments Map Explorer is viewed as a first step in GIS, with ArcView being bought once the skills, expectation, and demands of users surpasses the functionality offered by Map Explorer. Map Explorer has already been successful at opening up digital mapping and spatial analysis to more staff in the whole Council and has opened up the prospect of the public getting access to information the Council collects and holds.

The following diagram (figure 11.1) illustrates the relationship between the two technology solutions in the Council. Map Explorer is seen as 'GIS for the masses' with a large number of potential users and limited functionality.

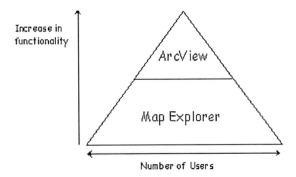

Figure 11.1 The relationship between two technology solutions in the Council

ArcView is the core GIS product with high functionality but a limited number of users.

Case study: educational services

An example of how ArcView and Map Explorer are delivering benefits to the Council can be found in the Council's Educational Services department. This department has always made intensive use of geographic information to support the running and operation of local schools and libraries and spatial information supports the diverse tasks of decision making, teaching, and public consultation. Any GIS implementation must take cognisance of how these tasks are performed and how the users want to use spatial information.

Within the core of the Council, geographic information is used by an educational support team for school catchment queries, pupil location queries and analysis of social trends such as educational attainment and social exclusion/inclusion. The GIS implementation for this team of one central ArcView seat supporting ten Map Explorer seats was based on the fact that the initial user requirements were to browse information that had already been collected. Map Explorer met these initial needs and now it allows staff to view the datasets held by the Council, and perform simple query and visualisation operations, producing simple thematic maps at the end of it all. Such tasks, however, would not be possible without ArcView, as Map Explorer is very limited with regards to data capture and data management/maintenance. The single ArcView seat has also been customised to the department's needs and provides additional education-specific functionality, powerful spatial analysis and query operations, and improved output options when compared with Map Explorer.

In this configuration Map Explorer is also acting as a stepping stone to ArcView – users are starting to become 'spatially enabled' and are developing both their skills and expectations of what the GIS can do. It is envisaged that once the demands placed on the GIS exceed the functionality offered by

Map Explorer, additional ArcView installations can more readily be justified and will be added.

The other application of spatial information in Educational Services has seen spatial information made available in schools and to the public. Opening up Council-held information is a key goal of the Council's corporate management team and currently it is GIS, and in particular Map Explorer, that is playing a lead role realising this aim.

In secondary schools, maps of the local area are used in the geography syllabus to teach students about map scales, map interpretation, population demographics, and so on. Unfortunately the maps held by the schools are old paper copies of OS mapping that have not only become very out of date but are beginning to fall apart.

The Council has recognised the potential of GIS to solve these problems while teaching students about the potential uses of geographic information, and has purchased PCs and colour printers for each of the nine geography departments in the local secondary schools. These PCs are being installed with Map Explorer and a selection of the key data sets held by the Council and give geography students the ability to query and visualise this information and include it in their own research projects. The decision to use Map Explorer was again based on an initial requirement to browse, visualise and print the Council's OS, environmental and socio-economic datasets and it is hoped that this access to Map Explorer will open up the potential for future applications of spatial information, perhaps involving ArcView.

The final application of GIS in Educational Services is in the main public library where GIS is being implemented in a public access-to-information project. While Map Explorer would seem like the logical solution for this, the need to tailor an application to the specific needs of the public coupled with the type of information being displayed meant that ArcView had to be used. The system had to be very simple to use and understand by the public and had to bring together many different datasets including scanned record drawings, a photographic library, and other digital archival material. Only ArcView allows for the development of a customised, wizard-driven application that meets these system requirements (see figures 11.2 and 11.3).

Summary

The decision by South Ayrshire Council to adopt a corporate approach to GIS implementation presented many problems, chief amongst these being how to deliver an appropriate technological solution to the desks of as many staff as possible. To solve this problem the different functionality and costs of ArcView and Map Explorer were exploited to open up the potential of spatial information processing within the organisation and make a failing GIS successful again. The implementation of Map Explorer has been particularly successful in unleashing the potential for GIS within the organisation and it

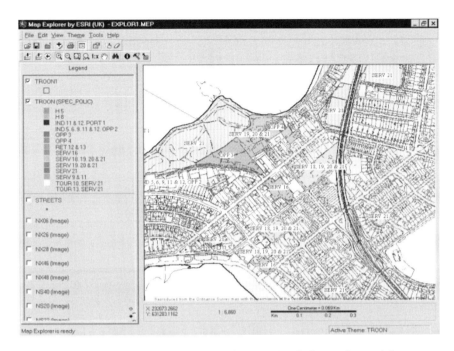

Figure 11.2 A screenshot from ArcExplorer (Source: Map © Crown Copyright)

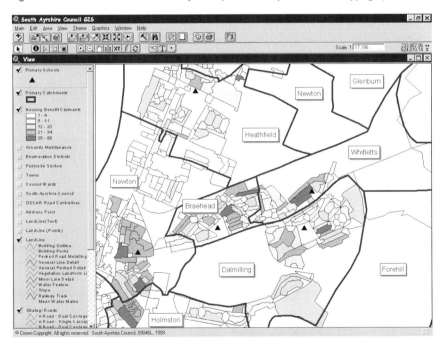

Figure 11.3 A screenshot from the South Ayrshire Council GIS

is now used at all levels to browse Council-held information. The result is that GIS in South Ayrshire in now no longer the domain of the well-trained operator or technician but has been opened up to all staff within the Council, school children and a wider public audience.

Chapter 12

Geography, GIS and the Internet

David R. Green

It was not all that long ago that the World Wide Web (WWW) or the Internet, as it is now often known, was a term with little meaning to all but a few people who were lucky enough to have access to the appropriate computer hardware and software technology. However, with time, public awareness and attention has grown, largely through the media (e.g. newspapers), and especially as access to faster computer systems in academia became the norm, and as universities in particular began to develop web sites, and to put together documentation packs, for staff and students alike, to post their own web pages onto the Net. Furthermore, over time, more web design and service provision companies have emerged, and there are now more Internet service providers, e.g. IFB (Internet for Business) and Demon, browser software providers such as Microsoft and Netscape, and telephone companies such as British Telecom (BT), taking an interest in the Internet. However, and this certainly appears to be the case in the UK, it is perhaps only in the last few years or so that many organisations and people are finally beginning to realise the importance of the Internet as a medium for rapid and efficient information delivery in both the home and the workplace.

The ever-decreasing costs of more and more powerful microcomputer hardware and software and a growing acceptance of the Internet technology as a means to communicate, to advertise, and even to shop twenty-four hours a day has also helped to make the Internet a more popular and familiar medium to more and more people. This has also been helped by the number of Internet providers offering introductory web and e-mail sign-up offers, free e-mail (e.g. Hotmail), and diskettes or CD-ROMs e.g. Compuserve and AOL (America Online) containing Internet browser software (e.g. Netscape Navigator, Microsoft Internet Explorer), trial periods, and software tools (e.g. HTML editors and image/graphics processing software).

However, whilst use of the Internet as a whole has been somewhat slower to gain acceptance in the UK (even in the commercial world) than the US for example, emphasis currently being placed on using the Internet in schools by the Labour government (see e.g. Carvel, 1998) is now beginning to help to further stress the importance of this technology within school education.

Rapid evolution of the different Internet software has also provided a whole new collection of tools enabling the rapid creation of web pages (e.g. Microsoft Frontpage and Macromedia Dreamweaver), and the delivery of both static and dynamic multimedia information (graphics, images, video and animation) via the Internet. Indeed anyone who has access to the Internet (at home or perhaps through an Internet café) via one of the main Internet browsers, Netscape Navigator or Communicator, or Microsoft's Internet Explorer, can soon begin to appreciate exactly how powerful a technology this really is. It is increasingly the case that many organisations, ranging from government, to academia, to commerce, and even individuals now have their own web pages (one or more) which can be accessed using the familiar uniform resource locator (URL) addresses e.g. http://www.abdn.ac.uk/. Indeed many schools now also have access to the Internet, as do many school pupils at both home and in school (e.g. Prudhoe Community School, Prudhoe, Northumberland, England) (see figure 12.1).

Whilst much of the information available via the Internet is not really geographical or spatial in origin, a great deal of it is in fact geographical and therefore of direct interest to geographers or in fact anyone who has a need for, or uses, geographical or spatial data and information. Maps, for example, are an important medium for communicating geographical or spatial information in the form of directions, locations and spatial patterns.

The graphics tools now available to web designers allow for the creation of simple static maps (e.g. in Paint packages, see Green and Calvert, 1998) that can be saved as .GIF or .JPG files, or screen captured using e.g. Paint Shop Pro™ 5 or Corel® Draw™ 4 software that are then directly Internet compatible (see figure 12.2). As such these raster images (images made up of pixels – picture

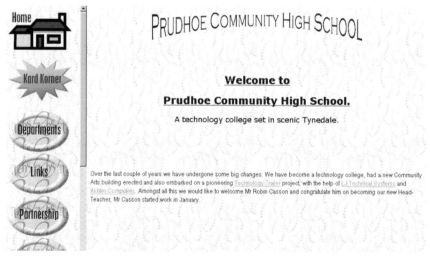

Figure 12.1 Prudhoe Community School website

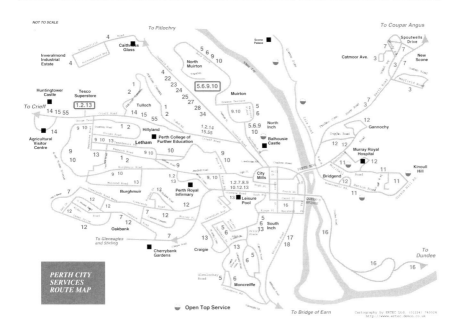

Figure 12.2 Bluebird Buses route map for Perth, Scotland (Corel® Draw™ 4)

elements) can be posted as part of Internet web pages, and even, if required, saved to disk on a local computer.

Static location maps

Many companies and organisations now post simple location maps (raster images) on their web pages providing customers with directions to their premises (e.g. http://www.adrian-smith-saab.co.uk/ is an example of a car dealership with garages in Aberdeen and Carlisle that provides location maps for people wishing to visit the showroom or for servicing) (see figure 12.3).

Dynamic maps

Extensions of these simple raster-based image maps can also be made interactive using software such as MapEdit (a software package designed to allow modifications to be made to the raster image in the form of 'hotspots' with 'hotlinks' to URLs) to create clickable image maps with clickable areas (figure 12.4 http://www.peyc.org.uk). Similarly, simple animated sequences of images (e.g. maps) can be generated using software GifTrans with the inclusion of various different effects or transitions between separate map frames (figure 12.5).

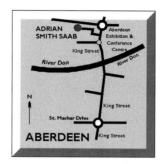

Figure 12.3 Internet location map for Adrian Smith Saab, Aberdeen

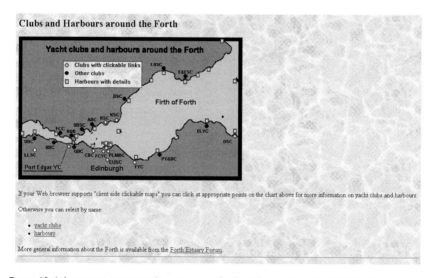

Figure 12.4 Internet interactive location map for Port Edgar Yacht Club (clickable hotspots) (Source: used with kind permission of PEYC, and taken from www.peyc.org.uk)

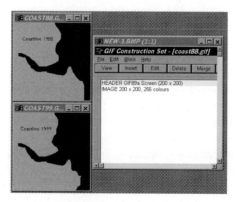

Figure 12.5 Simple animation using GifTrans software

Interactive mapping

Maps (or indeed any text or image files) which originate as Adobe® Acrobat® .PDF files can also be spawned to the Acrobat Reader (e.g. Acrobat® Reader™ 3.0 which is available from the Adobe® website http://www.adobe.com/), and using the Reader tools, the user can zoom and pan the map. An example of the use of .PDF files can be found on the Forth Estuary Forum (FEF) GeoInformation website (http://www.abdn.ac.uk/fef/) showing maps of the Firth of Forth (figure 12.6). The London Transport webpage also provides access to .PDF format maps e.g. (http://www.londontransport.co.uk) (see figure 12.7).

Other examples

There are, of course, many other examples of spatial data and information that can be posted to the Internet, including aerial photography and satellite imagery, all of which can be used as described above. Many map and image data catalogues, for example, use an image map as the basis for user-navigation. Simply by pointing and clicking on the map hotspots, the user is linked to another image or map on another web page (see figure 12.8).

Internet

GIS technology

But the Internet technology has, in the past few years, developed far beyond the simple delivery of static, interactive or animated maps created by using the above-mentioned software. Nearly all GIS software developers (e.g. ESRI [http://www.esri.com/], Autodesk [http://www.autodesk.com/], GeoConcept [http://www.geoconcept.com], Intergraph [http://www.intergraph.com/]) have developed the capability to deliver maps, including interactive maps over the web, and 'on the fly'. Whilst many examples of these map servers use raster maps, some are now using vector technology e.g. GeoMedia from Intergraph (http://www.intergraph.com).

ESRI, amongst many others, has developed a range of products (e.g. ArcIms) designed to enable users of their GIS products to deliver maps created using Arc/Info and ArcView over the Internet using a similar interface to ArcView which allows maps to be created 'on the fly' and to toggle so-called map layers, themes or coverages on and off. The GIS tools also allow the end user to zoom in and out of the maps, and to pan around the maps, as well as some additional simple functionality. A very good example of the potential of the Internet as a means for delivering maps and spatial information is the Florida Online Mapping system (http://www.fmri.usf.edu/sori/Default.htm) (figure 12.9). (Note: this particular site is Java-enabled and requires a fairly

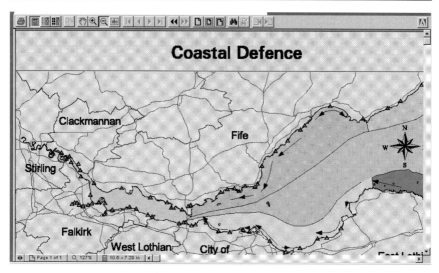

Figure 12.6 Forth Estuary Forum GIS map in portable document file (pdf) format (Source: based on information derived from and copyright of Bartholomew 1 : 250,000 GB database, Scottish Natural Heritage, RSPB Estuaries Inventory Database. Produced from Forth Estuary Forum GIS © Forth Estuary Forum, 1998)

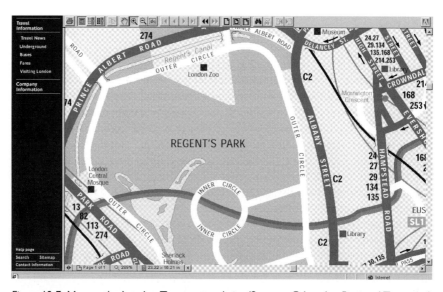

Figure 12.7 Map on the London Transport website (Source: © London Regional Transport)

well-specified PC platform with a fast processor, plenty of RAM (random access memory), a good graphics card with on-board RAM, and a fast modem or connection to the SuperJanet network or an ISDN line to provide successful and rapid access to the maps.)

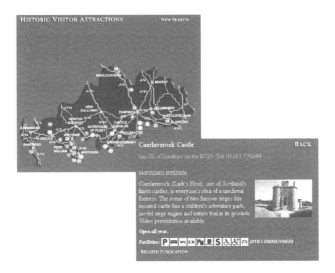

Figure 12.8 Linking images via hotspots (Source: © Historic Scotland)

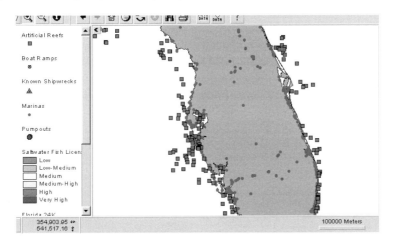

Figure 12.9 Florida Online mapping system (Source: © Florida Fish and Wildlife Conservation Commission – Florida Marine Research Institute)

There are of course many other good examples on the WWW which similarly demonstrate the power of the Internet and associated technology to deliver spatial data and information, much of it in the form of maps. In the US, some of the sites also provide a means to access spatial databases, and even to add a 'value-added' product back into the database for the benefit of others using the 'database'. This practice is not, to the best of my knowledge, yet possible in the UK.

Map catalogues

Another example of the value of the Internet to end-users is the provision of access to map catalogues (including metadata – that is information about information), and online mapping systems. Whilst not falling into the category of a true GIS, such systems nevertheless provide access to a form of GIS which involves simply delivering raster-based maps to the end-user based upon navigation via an image map, or via a search using a place name or a postcode.

Postcode location

In some ways searches for car dealerships, for example, are primarily geographical in the sense that the user inputs a postal code or town (a locator address), and then the search tool retrieves the dealership at the location matching the input address e.g. Saab car dealerships (accessible on the Saab site at www.saab.co.uk) (figure 12.10). There are many other similar examples.

Information systems

One of the many uses of GIS, albeit in a somewhat simpler form than an analytical toolbox, is as an information system, whereby maps are the main form of output to the end-user. Besides simply delivering maps at the click of a button, a number of Internet-based public information systems make widespread use of maps: for example, in Toronto, Ontario, Canada (http://www.city.toronto.on.ca/ttc/schedules/index.htm) (figure 12.11). There are

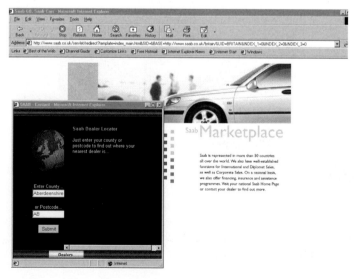

Figure 12.10 Postcode nearest dealer search on the www.saab.co.uk website

Figure 12.11 The structure and operation of the Toronto Transit Commission (TTC) (Source: © Toronto Transit Commission)

Figure 12.12 The Solway Firth Partnership (SFP) website: an example of an internet-based public information system

many other examples of the use of the Internet as an information system (figure 12.12).

Access to the Internet

All these examples, and many more, are readily available over the Internet right now. Whilst not all schools (elementary, primary or secondary) are well endowed with good computing facilities yet, or links to the Internet, this will hopefully change with the recent announcement of funding by the Labour government mentioned at the start of this chapter. Increasingly, the Internet is becoming accessible at home, although the numbers of homes with access to e-mail and the Internet is not yet perhaps as widespread in the UK as it is in the US. The advent of digital television may of course begin to change all this when the Internet becomes accessible via the home TV set. It is, however, highly likely that the Internet will soon become as popular as pagers and mobile phones as the technology becomes more affordable, easier to use, and people begin to both appreciate and realise the potential of the Internet as both an information resource and a communication system.

Placing static and simple interactive maps on a Web page

Although not all that difficult to master, the creation of web pages requires at the very least mastering of HTML (the hypertext markup language), some simple image-processing software (e.g. Paint Shop Pro™, Lview/Lview Pro, Pictureman or Corel® PHOTO PAINT®), a scanner (preferably colour) with both a graphic and an OCR (optical character recognition) capability and, in addition, if a little more advanced, some knowledge of and software for VRML (virtual reality markup language), Netscape's JavaScript, and Sun's Java, as well as access to a digital camera (640 × 480 pixel resolution and higher).

Today the potential web designer is also aided by the availability of a number of software toolboxes that allow virtually anyone to create web pages relatively easily without a great deal of HTML knowledge. This has also been helped by the growing availability of cheaper hardware and software and special software packages such as Microsoft's Frontpage which offer point-and-click tools with menus to construct the HTML-encoded web pages. These can be viewed and edited prior to posting on the Internet. Word-processing files created in Microsoft Word can now be saved as HTML files which can then be directly posted including all the styles such as bold, underline, colour and so on (although not all attributes such as coloured horizontal lines will show in the Netscape browser). Similarly, slides and slide shows created in Microsoft PowerPoint can be saved to HTML files including the text, graphics, charts and tables, and 'played' using the 'buttons'

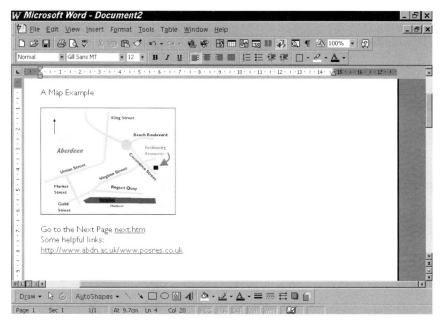

Figure 12.13 Microsoft Word text file (prior to conversion to HTML file)

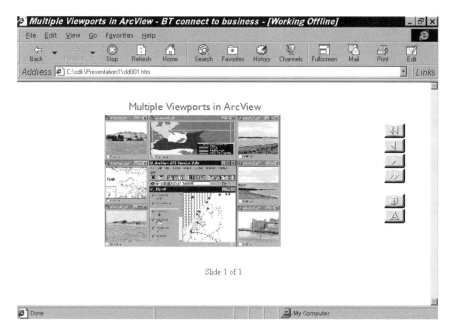

Figure 12.14 Microsoft PowerPoint presentation for the Internet (HTML file) (Graphic image supplied courtesy of Environmental Systems Research, Inc. and is used herein with permission)

Table 12.1 File transfer protocol (FTP) general use commands

FTP command	Action
close	ending an FTP session
quit	quitting FTP
prompt	toggling the prompt off
get	retrieving a file
mget	multiple file retrieval
mput	multiple file sending

added to the HTML code. Some examples are a Microsoft Word text file (see figure 12.13) and a Microsoft PowerPoint presentation (figure 12.14).

A little knowledge of FTP (file transfer protocol) software is also useful e.g. WS-FTP (often available from cover CD-ROMs on computer magazines), PC-NFS, and using some sort of UNIX interface e.g. TelNet software. FTP allows you to move files to and from a local PC and a remote server. Both text (ASCII) and image/executable files (binary) can be transferred. FTP is the usual way that you would post web-page files to webspace on a web service-provider's server. An example follows:

```
TelNet
Login
Password
sysa% ftp
sysa% open www.netx.co.uk
Userid>
Password>
ftp> bin
ftp> put image.gif
```

In other words after logging on via e.g. TelNet, FTP is started, followed by setting the file transfer type to BINARY, and sending a .GIF image file from the local PC to a remote server.

Other FTP commands of general use are shown in table 12.1.

The creation of an Internet page is relatively easy to achieve. The following is an example.

If a package such as Frontpage is not available, it is relatively easy to create the HTML code using Microsoft's Wordpad, saving the end result as a .htm text (ASCII) file.

```
<html>
<title>A Geographical Example</title>
<meta>
</meta>
```

```
<body bgcolor="#ffffff" text="#000000>
<p>
<font face="arial" size="3">
<b>A Map Example</b></font>
<p>
<hr>
<p>
<center>
<img src="map.gif">
</center>
<p>
<hr>
<a href="next.htm">Go to the Next Page</a>
<hr>
Some helpful links:
<li><a href="http://www.posres.co.uk>www.posres.co.uk</a>
</body>
</html>
```

Where   <p> = paragraph; <hr> = horizontal line; <li> = bullet point; <body> = main body of web-page code; <html> = html page; <img src> = image; <center> = center whatever follows; <a href> = link to another page on the same website or to another (http://); <font> = specify font type and size etc.; <title> = title of the webpage; bgcolor = background colour (specified as RGB – 000000 = black ffffff = white).

In most cases the coding starts and ends, where the end of the code is denoted by a forward slash e.g. <center> followed by a </center>.

This is only a simple example, that does the following:

• Displays a line of text (ARIAL font; size 3; left justified)
• Horizontal line
• A centred image
• A link to another (hypothetical page)
• A link to another website
• Is spaced out using paragraph breaks

The output is shown in figure 12.15.

The file map.gif (a Compuserve graphics image file format) (or it could have been a JPEG – joint photographic expert group file format, e.g. map.jpg), containing a simple map drawn in a 'graphics design' or 'paint package' (see Green and Calvert, 1998) is easy to create. Software packages such as Paint Shop Pro™, Pictureman, Corel® PHOTOPAINT® or Corel® Draw™ can be used to create the graphic, either saving or exporting the result to a .GIF or .JPG file.

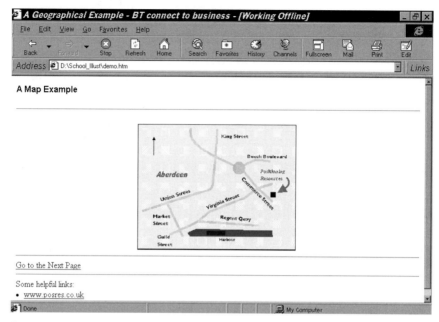

Figure 12.15   HTML file viewed in the Microsoft Internet Explorer 4 browser

More complex examples can be constructed using the HTML language, VRML, JavaScript and Java applets. It is best to consult either a textbook on HTML and web-page construction, and there are many publications now available in this burgeoning marketplace, or some simple tutorials that offer a starting point (e.g. Bell and Parr, 1998). Additional help and suggestions can be found by searching the WWW using one or more of the search engines e.g. Lycos, AltaVista or Yahoo using keywords such as HTML, web pages etc. Many university sites also provide instructions and tutorials on the creation of web pages, and some include downloadable web construction tools, together with buttons, page backgrounds, and other graphics. Some useful website URLs are listed at the end of this chapter.

Summary and conclusions

The use of the Internet in education, commerce and government as an information resource is growing very rapidly. Tools to facilitate the creation of web pages have become more widely available (often as freebies on computer magazine covers, or as shareware), and as the computer hardware and software becomes cheaper more individuals are both creating web pages and using the Internet at home and in the workplace. Schools are one place where both teachers and pupils now have access to the web, and in a fair number of cases, schools are also mounting their own websites to enable them to communicate locally, nationally and internationally.

As GIS technology on the Internet becomes more commonplace, so it will be possible to make more use of the web in the classroom for geographical studies, primarily as an information resource. Already it is possible to see many examples of geographic information on the Internet, ranging from simple maps, through to online mapping tools, all of which can be used online, downloaded and placed in documents, or interacted with directly. The Florida Online Mapping system is a particularly good example. Other useful examples, of direct value to schools, are the website and GIS materials available via ESRI (http://www.esri.com). Another useful illustration is provided by the Friends of the Earth site (http://www.foe.org.uk).

It would seem that in the future much can be gained from Internet access for GIS in a school environment. The Internet provides a simple means to deliver and access information, as well as to communicate and to gather information. The ESRI site provides an excellent example of how the Internet can be used in the context of education as follows:

- Provision of access to data and information in the form of data, text, images, and maps
- Provision of access to software (freeware, shareware, and to purchase downloads or via Internet shopping)
- Provision of access to online and downloadable teaching materials (paper-based and electronic publications and tutorials, interactive maps and GIS, and videos)
- Provision of up-to-date news and information
- Provision of a means to communicate with fellow teachers/pupils and to take advantage of additional resources
- Provision of access to links
- Provision of links with software and data suppliers

The ESRI site also offers clear insight into how far the technology has come in only a few years. Whilst many other sources of information about GIS are available in the form of books (such as this one), demonstration and other software on diskette and CD-ROMs (from companies and organisations), publications (e.g. from the AGI), and theses, the Internet offers a very powerful medium to use as a resource for both teachers and pupils alike, and to explore learning about and using GIS in an increasingly geographically enlightened world.

References

Bell, D. and Parr, M. (1998). *Java for Students*, Prentice Hall Europe, 570p.
Carvel, J. (1998). 'Blairs' £1bn IT Package for Schools', *The Guardian*, Home News Section, Saturday November 7, p. 7.

Green, D.R. and Calvert, L.P. (1998). 'The Cartographic Potential of Graphics Design Software for Education', *The Cartographic Journal*, Vol. 35(2), pp. 61–70.

Further reading and information

Graham, I.S. (1996). *HTML Sourcebook: A Complete Guide to HTML 3.0*, Wiley & Sons, New York, 688p.

Lemay, L. and Perkins, C.L. (1996). *Teach Yourself JAVA in 21 Days*, Sams.net, Indianapolis, IN, 527p.

Walsh, A.E. (1996). *Foundations of Java Programming for the World Wide Web*, IDG Books, Foster City, CA, 906p.

Chapter 13

Tailoring GIS courses for employment

Michael Gould

The changing workplace

GIS curriculum design has been the topic of many papers and conference sessions during the past ten years (for example, Kemp and Dodson, 1991, list eighty papers), and emphasis has been placed on the categorisation of students (future users) and course formats. Many of these works define a comprehensive GIS course to serve all possible students: a sort of minimum requirement of GIS 'vitamins' in a general educational diet. Unfortunately, many future GIS users will not be specialists, and will not attend traditional university courses or the other requisite courses – statistics, computer science, management, geographic problem solving, cartography, etc. – which all form a part of any comprehensive twelve to forty-eight month graduate degree programme in GIS. Is it reasonable to expect that the hundreds of thousands of users of Microsoft Office (which includes a subset of MapInfo) will invest in an MSc in GIS or even that they had attended a GIS course during their college days? The rapid evolution of GIS ease-of-use and functionality allows that users need not make this investment. Those few who wish to become GIS developers are a special case.

Question: How do people become GIS users? Answer: Depends on where they live. A survey by *GIS World* magazine (March 1995) shows that two-thirds of GIS users in certain fields (natural resources and environmental industries) are in fact self-taught, half of all GIS users received training from short courses, a quarter have graduate degrees and 18 percent have under-graduate degrees related to GIS. We must be careful in extrapolating these results, however, because they reflect the magazine's North American bias. For example, many European countries have not had graduate or under-graduate GIS programs in place long enough to have produced any graduates yet! The figures for Spain, then, would be closer to 80 percent self-taught, 15 percent from vendor or other short-course training and 5 percent from traditional university curricula. On the surface it seems obvious that the immediate priority should be to implement comprehensive GIS curricula in Spanish universities to produce more graduates with a well-rounded view of GIS. After all, the British and the Dutch have done so. However, while in

Spain we expected to see a quadrupling of the number of GIS users who came straight from university programs by 1998, the key word is *priority*. We must ask ourselves if – in the midst of economic crisis and 20 percent unemployment – the long-term goals of the comprehensive GIS curriculum ought to have priority over the immediacy of putting people to work; or, what type of GIS user should we be producing today? As far as the supply side of the question is concerned, the students themselves have adapted to what they believe to be the answer.

We have noted a curious trend during the past few years in Spain, regarding the short courses offered at the Universities Complutense de Madrid and Politecnica de Madrid. These courses are open to the public, offer training in specific GIS software (ArcCAD and PC Arc/Info), cost between US $300– 800, and run from 20 to 40 hours each. Three years ago the great majority of students were government employees and commercial professionals seeking retraining or specialisation within their current field. Today we find that the majority are advanced graduate students and unemployed graduates, from a wide range of fields (we should also note that when we offered a general course called 'Introduction to GIS' few people registered, whereas adding Arc/ Info to the title brought much interest). This shift in student composition is due to economic factors: government agencies have cut back on extramural activities and matriculating students are finding that four- or five-year degrees are not getting them employed. These short-course students (especially the unemployed ones) are acutely aware of what today's job market demands: the niche worker rather than the Renaissance man (sic).

Let us look at the demand side of the situation. In Spain, as in other nations, GIS use is growing both horizontally (entering more diverse fields) and vertically (more users in each field). However, to see where the jobs are we must examine the GIS industry at higher resolution, that is at the level of single employer behaviour. There is a growing trend among employers using GIS, which is also directly related to the current economic crisis: implementation of pilot projects. Almost all new GIS implementations, in small, medium, and large companies and institutions, are so-called pilots, which provide several comforting (although often fallacious) characteristics. Pilots are safe because they are brief and, thus, relatively immune to change of government or chief executive officer (CEO). Brevity also makes pilots (appear) relatively cheap: fewer person-hours, fewer machines etc. means fewer ecus spent. Pilots also provide a perfect excuse to hire workers on a non-contractual, temporary basis. In sum, pilots offer a great deal of flexibility during an unstable economic period. Pilots also have several negative characteristics which may adversely affect the long-term system design and implementation strategy, but that topic is beyond the scope of this chapter.

What does the trend toward pilot projects mean to the student of GIS? For the young student just beginning his/her university education it is of little importance, because the trend may change several times before graduation in three to five years. For the advanced student, about to begin the job search, it will likely cause hardship as many new GIS user-organisations are implementing their pilot GIS on a one-machine/one-technician basis. (Note: *GIS World* announces that when North American government agencies award contracts, they are for 500 or 1,000 'seats' or copies of software at a time. In Spain an entire ministry might buy two or three). Based on this short-term and conservative implementation model, the employer is searching for niche workers: one to operate the database management system (DBMS) and another couple to digitise 100 maps, for example. And employers running pilot projects are not hiring middle managers, who demand too high a salary and a stable contract.

This is contrary to opinions of prominent North American employers (in Gilmartin and Cowen, 1991), that graduates with overall knowledge in a related cognate field (i.e. biology, geography, planning) and a certain level of technical skills – not hyperspecialists – have the distinct advantage in the workplace. Again we cannot extrapolate to other countries: these opinions presuppose a healthy economy and abundant employment opportunity, and that projects are long-term investments which evolve over the period of five to ten years, permitting 'on-the-job' training. The situation in 1995 in much of Europe, however, was quite different. Pilot projects (at the time of writing) demand that each worker demonstrates immediate productivity. Where does the recent graduate of a five-year university program and with general GIS knowledge fit in? He/she is not (normally) an experienced DBMS user or UNIX system operator and, on the other hand, does not wish to spend a year digitising at near minimum wage. This latter problem is very real in Spain today, and it is disheartening watching competent university graduates reluctantly accept temporary employment as interns (at low or no wage). Hiring of interns is on the rise due to the Spanish government's recent incentives encouraging employers to do so. Thus, it is made to appear that unemployment is decreasing, whilst we all know that these interns will be on the street again in six months.

A key dimension: education versus training

Separating basic education from practical training has been a common thread running through 'GIS in Education' (now Education and Awareness Special Interest Conference Committee [EASICC]) sessions of past conferences. We constantly debate what the primary role of the university should be: to provide young people with basic education or more directly to increase their value in the job market. Under the best of economic conditions it is possible to ignore the latter and focus on basic education. Not surprisingly, then, North Americans usually insist that practical training should not enter into the university picture. Let the vendors supply specific training, they say. Again, assuming that GIS users are able to pay US$1,500 plus expenses to attend vendor short courses is only reasonable under favourable economic conditions. While we do not believe it is wise to ignore the university's fundamental goal – basic education – in countries such as Spain we must also attend to localised situations such as the economic crisis at the time of writing and assist in any way we can. This involves providing specialised training, aimed at filling various niches in the job market.

Many academic departments have recently begun offering short GIS training courses, directed mostly at working professionals but also serving the unemployed (or the underemployed). While serving the community and modestly helping to ease the unemployment situation is commendable, the main reason for running these courses is the economic survival of the department, which otherwise has little budget to support a GIS/remote sensing laboratory for the use of its own students. Running short courses for public consumption is not extraordinarily profitable, but it does underwrite part of the cost of hardware and software maintenance, helps to fund graduate students (which in turn helps keep hold of them until they complete their studies), and allows the department to sponsor visiting lecturers. As one example, a thirty-hour GIS course taught at the Universidad Complutense de Madrid generates approximately 1,200,000 pesetas (US$9,300) of which about half goes towards professor honoraria and per diem, a third or more directly to the department for maintenance and amortisation of the GIS laboratory, and the rest for general operating costs (photocopies, publicity, the university foundation's 5 percent). Thus, offering two such courses per semester allows for the purchase of a few pieces of hardware (printers or digitising tablets for example) and/or the maintenance of the laboratory's software. Furthermore, the residual effects of these short courses – teaching experience for graduate students, industry contacts (which often lead to collaboration), and the possibility of filling vacant seats with university colleagues (who get a free ride) – do not go unnoticed.

The emergence of these university training courses affects both basic education and professional training institutions as well. Because the same university professors teach both types of courses, lessons learnt during the intensive short courses naturally find their way into academic courses at the undergraduate and graduate level. It suddenly becomes easy to offer laboratory sessions (recently tested) in a course which historically had been predominantly theoretical. Furthermore, professional students' experiences and comments – gleaned from the short courses – can be transmitted to undergraduates, adding a dimension of reality to the course. These new training courses may also enter the terrain of state-supported professional training schools. This is the case in Spain, which has recently reformed its education system to accommodate two levels of practical training (Formacion Profesional, or FP): state schools, which offer non-university diplomas in computing, CAD, business studies, etc. Figure 13.1 shows the recently modified Spanish educational tree, which provides two routes towards a job market that demands technical (practical) training and university education.

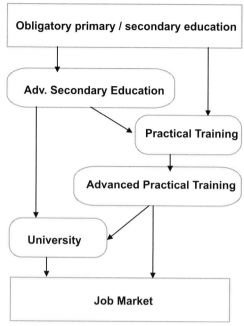

Figure 13.1 The recently modified Spanish educational tree, which provides two routes towards a job market

The reform is recent, and we have yet to see any FP institutions offer GIS training, but this advanced level seems an ideal home for courses on how to operate GIS software: from base data compilation, to automation, to production via macro language programming. A strong offering at this FP level would free the universities to get back to the business of basic (theoretical) education. But is this what the universities want? They would lose their grip on this trendy technology which helps so much to bring in grant money and provide a steady flow of toys to play with! As the GIS field evolves into millions (not just thousands) of users, perhaps we will need to consider – again – both basic education and practical training, and who should provide what to whom.

Another key dimension: informatics versus thematics

'There are two types of people in this world: those who divide people into two types, and those who do not' (Anonymous). While GIS specialists come from a wide variety of fields, in nations like Spain the categorisation is more closely binary: the 'techies' and the 'non-techies'. Let us look at an example of this divide in disciplinary terms. The Environmental Agency of the region of Andalucia (southern Spain) has one of the best developed GIS/remote sensing departments in the country (see Moreira, Gimenez-Azcarate and Gould, 1994, for details). It employs twenty-seven non-managerial workers, divided into two mutually exclusive groups: 'informaticos' (digital information specialists) and 'tematicos' (experts in the cognate environmental sciences). Each project is handled by small teams composed of at least one worker from each group. This mix, according to the department's director, guarantees that each client's thematic and informatic needs will be recognised and met. The tematico (biologist, geographer, geologist, etc.) speaks the client's real-world language, and the informatico knows how to translate needs into information products.

This dichotomy provides an ideal working situation in a nation like Spain, where interdisciplinary university study is rare: if it is difficult to create a well-rounded GIS expert, at least one can form an interdisciplinary team or talent pool. But let us look at the GIS education/training side of the picture. To meet the needs of this department, the education community must essentially teach GIS at two levels: GIS for number crunchers and GIS for scientific-users. The 250-hour course offered by the Universidad Complutense (see table 13.1) is clearly oriented towards the latter: only a passing reference to UNIX, no C++, no device drivers, no DBMS programming. Approximately 95 percent of the people who solicited the course had no substantive programming experience. Some people in the GIS community argue that any respectable GIS course must focus on technical aspects – after all we are talking about an information system. But the organisers of this course, as geographers, feel that it is equally necessary to understand the basics: the theoretical framework and the possible application of GIS to real-world problems. Thus, we counter-argue that a person does not get very far treating coastal erosion or non-point source pollution as a string of bytes! But let us not confuse a non-technical approach towards teaching GIS with what Kemp et al. (1992: 183) argue: 'an emphasis on education is more compatible with the objectives of a university and would be more appropriate for students planning to concentrate on applications and research'. Their figure 1 (Kemp et al., 1992: 184), showing two dimensions of GIS specialties, is again assuming abundant job opportunities whereby the student merely selects his/

Figure 13.2 Two axes forming four possible GIS course types

Table 13.1 The curriculum of the GIS short course at the Universidad Complutense de Madrid

Module I – Introduction to GIS (40 hours)
- Definition of GIS
- Cartographic basics
- Typology of GIS and of databases
- Basic demonstration with Atlas GIS
- Applications of GIS
- Review of software on the market

Module II – Raster-based GIS (30 hours): practicals using IDRISI
- Fundamentals of the raster data model
- Typical data formats
- Data loading: various alternatives
- Operations (local, neighbourhood, zonal)

Module III – Vector-based GIS (80 hours): practicals using PC Arc/Info
- Fundamentals of vector data models
- Data loading
- The digital map: geometry and topology
- Building the attribute database
- Spatial and logical queries
- Buffer and overlay analysis
- Network analysis
- Cartographic composition and output
- Basic macro language programming (SML)

Module IV – Remote sensing (40 hours): practicals using ERDAS™ (PC)
- Basic concepts: electromagnetic spectrum, resolution, distortion, etc.
- Sensor types and their characteristics
- Data management
- Visual and digital interpretation of images
- Integration with GIS

Module V – Practical Project (50+ hours)
- Class of sixteen divided into partners
- Each group works on the same project: optimal corridor determination
- Each group able to use any software available
- Graduate students always present to assist

Module VI – Wrap-up (10 hours)
- Presentation/discussion of final projects
- Employment counselling (Unemployment Agency personnel)

her career path, attends the necessary course(s), and then begins working. In Spain, lack of funding means there is practically zero market for PhDs, either as professors or as researchers. This fact wipes clean the upper right of Kemp et al.'s graph. And since GIS development is practically nil in Spain, the lower right is gone as well. Now all levels of Spanish education and training institutions are essentially battling for what remains of the middle and left. Perhaps the graphics in Figure 13.2 makes more sense in a non-North-American context. In this figure, two key axes form four possible GIS course types. Perhaps shifting attention from one to another course type would allow the flexibility to meet the changing needs of the job market.

Call for flexibility

Instead of providing a 'field' of many possible combinations of specialties as in Kemp et al. (1992) or Toppen (1990), figure 13.2 provides a much more discrete division into four GIS courses. In certain countries only a few of the four will be possible or necessary, while other nations have sufficient demand for all. In the case of Spain, there is little demand for course type 4 (basic GIS education for computer scientists) due to the lack of research positions and system design/development. The majority of the university GIS courses in Spain are currently type 2 (basic, theoretic courses for biologists, geographers, agricultural engineers, etc.). The recent surge of university short courses (such as in table 13.1) fills the type 1 niche (teaching biologists or geographers how to use GIS packages). Type 3 courses (applications development using macro languages) are desperately needed in Spain, but perhaps should be taught only by vendors or the practical schools. The ability to emphasise certain courses over others should be facilitated so that in times of crisis courses 1 and 3 may be stressed (and better funded), while courses 2 and 4 may be stressed as the clouds begin to clear. Focusing only on traditional university education means that we necessarily suffer from a three- to five-year lag in our reaction to employment needs. Are we constantly out of phase?

Table 13.1 shows the curriculum of the 250-hour (four-month) course offered by the Universidad Complutense de Madrid and funded by the national unemployment agency. The course was without cost to the students, of which sixteen were admitted from approximately eighty solicitations. Course methodology was approximately 80 percent practical, 20 percent lecture. Note that the course is of type 1 according to figure 13.2. Plans are to expand the course to 500 hours (a 'master's' degree) and, thus, have time to focus on more theoretic topics (course type 2).

References

Gilmartin, P. and Cowen, D. (1991). 'Educational Essentials for Today's and Tomorrow's Jobs in Cartography and Geographic Information Systems', *Cartography and Geographic Information Systems*, Vol. 18(4), pp. 262–7.

Kemp, K.K. and Dodson, R.F. (1991). *Bibliography on Teaching GIS*, University of California, Santa Barbara: National Center for Geographic Information and Analysis, unpublished photocopy, 6 pp.

Kemp, K.K., Goodchild, M.F., and Dodson, R.F. (1992). 'Teaching GIS in Geography', *The Professional Geographer*, Vol. 44(2), pp. 181–91.

Moreira, J.M., Gimenez-Azcarate, F. and Gould, M. (1994). 'Evolution of an Environmental Information System', *GIS World*, Vol. 7(11), pp. 46–9.

Toppen, F. (1990). 'GIS Education in the Netherlands: Bits Everything and Everything About a Bit?', *Cartographica*, Vol. 28(3), pp. 1–9.

Chapter 14

GIS education in Germany

A survey and some comments

Ralf Bill

Some years ago the author prepared a first analysis on the situation at German Universities concerning GIS education (Bill, 1992) followed by an inter-disciplinary workshop on GIS in education (IFP, 1992). At that time a very limited number of institutions offered relevant GIS courses regularly. In 1996, this survey was again carried out by the author revealing that the situation had changed completely. By then, a broad range of GIS-related lecture units were offered at German schools in a variety of scientific disciplines at all levels in education.

About 500 institutions at German-speaking universities and polytechnic schools (FHS), covering nearly all of the institutes possibly teaching GIS-related topics, were asked to answer a four-page questionnaire on their situation with respect to GIS in a broad sense. About ninety institutions including a broad range of disciplines replied to this request. These disciplines were arranged into the following major groups (given in alphabetical order) which were then investigated and presented in this chapter (with, in brackets, the number of replies related to study courses at universities/FHS):

- Agricultural sciences (6/1) including agricultural ecology
- Architecture, civil engineering and spatial planning (3/2)
- Computer sciences (4/-)
- Economic sciences (2/1) including computer science in economy and economic geography
- Environmental science (7/1) including environmental monitoring, environmental engineering, land cultivation and environmental protection
- Forestry (1/2)
- Geodesy (1/10) including surveying and photogrammetry, geoinformatics and cartography
- Geography (27/-) which summarises physical and cultural geography as well as geo-ecology, geology and others
- Landscape planning (3/6) including landscape architecture and land cultivation

The number of replies differs for the groups. However, the sample seems to be sufficient to derive some of the major characteristics. The number may also reflect the importance different faculties assign to the topic of GIS. Figure 14.1 presents the spatial distribution over Germany and the subjects taught at these places.

The situation in GIS education

Some general statements

The data from the questionnaire were stored and analysed in a Microsoft Access database making use of a user-friendly interface form (figure 14.2). The statistical evaluation of the data include some general information on the faculties, the number of students, the final academic degree and so on. The study courses lead to qualifications in a diversity of professional careers such as geodesy, cartography, soil science, geology, geography, civil engineering, architecture, computer science etc. Usually the final degree is the diploma, in some cases the schoolmaster, reached theoretically after four years of study. The number of students per year differs considerably between the disciplines.

In total, around 22 percent of the students covered in this survey (ca. 13,000) do have the chance to get an education in GIS-related subjects, but with very different proportions comparing the individual sites and courses of study. At universities more option subjects are provided than compulsory ones, whereas in the FHS the opposite is the case. This is valid in general for lectures, exercises, seminars and laboratory exercises. At both school levels (university and FHS) more lectures are given compared to exercises and a much smaller number of seminars and laboratory exercises. At FHS about 50 percent, and universities around 38 percent of the GIS-related contents are offered in a lecture form (see table 14.1).

In principal the basic disciplines teaching GIS in Germany are geography and geodesy (together more than 50 percent). Because of the small number of students in geodesy compared to geography, the major contribution to the GIS job market comes from geography. All other disciplines are preparing fewer students and devoting a smaller proportion of lectures during studies for the GIS market.

German educators also prefer German textbooks. Currently three German textbooks on GIS are available. A strong preference is given to the textbook by Bill and Fritsch (more than 60 percent) followed by Bartelme (50 percent) and Gopfert (44 percent). Looking at English-based lecture material, Burrough's book scores 33 percent and is the most popular text followed by books by Aronoff, Star and Estes, whilst Tomlin and Laurini and Thompson are less important (14 percent). Educators combine textbooks (62 percent) with their own material (62 percent) and product manuals (48 percent) to

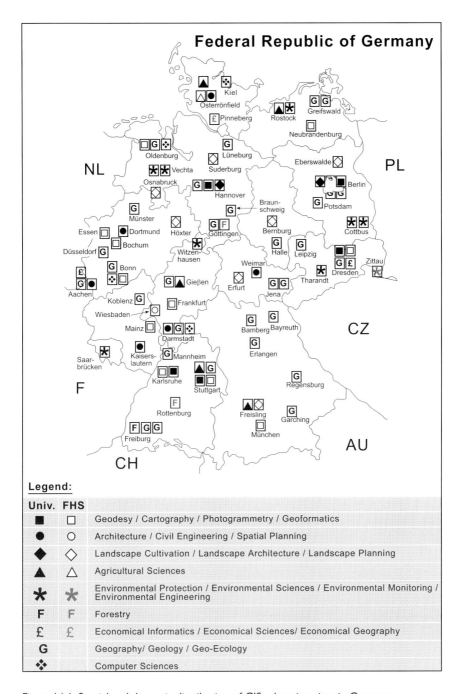

Figure 14.1 Spatial and thematic distribution of GIS education sites in Germany

Table 14.1 Percentage of students and number of GIS hours in the different disciplines

Discipline	Students in GIS course (%)	Hours on GIS	Semester
Agricultural sciences	11	1–6	5–6
Architecture	32	6–10	1–6
Computer science	28	2	6
Economic science	3	1–4	1
Environmental science	53	1–36	1–8
Forestry	8	0–9	7
Geodesy	53	2–31	2–7
Geography	18	2–36	1–7
Landscape planning	38	3–10	2–5

create their own lectures. The NCGIA core curriculum, GIS Tutor 2, GIS product tutorials and lecture materials available over the Internet find less acceptance in the education community.

Lectures by discipline

The following is a list of lectures named in the individual disciplines together with some comments on the relationship between obligate and optional subjects ranging from lectures to laboratory exercises.

Agricultural sciences   Fundamentals on GIS are offered, in some cases in other courses on soil science, soil informatics, hydrology or as practical laboratory exercises with a smaller proportion on GIS. In many cases GIS is an optional subject.

Architecture, civil engineering and spatial planning   GIS is often presented in standard computer science lectures, as part of utility lectures or together with planning techniques or CAD. In spatial planning an equal number of lectures and exercises exist. GIS is taught as a compulsory subject and sometimes as an optional subject in the form of seminars. In total, the number of GIS-relevant topics is not very closely related to other disciplines.

Computer sciences   GIS is still the exception, offered only as an optional subject in lectures with fewer exercises or via seminars and practical laboratory courses. GIS lectures on information systems in general (with small parts on GIS such as temporal and spatial aspects), environmental information systems, database systems, and GIS.

Economic sciences   GIS is still an exception. In Dresden, GIS 1 and 2 are delivered, whilst at other places GIS is offered in tutorials as optional components of the studies.

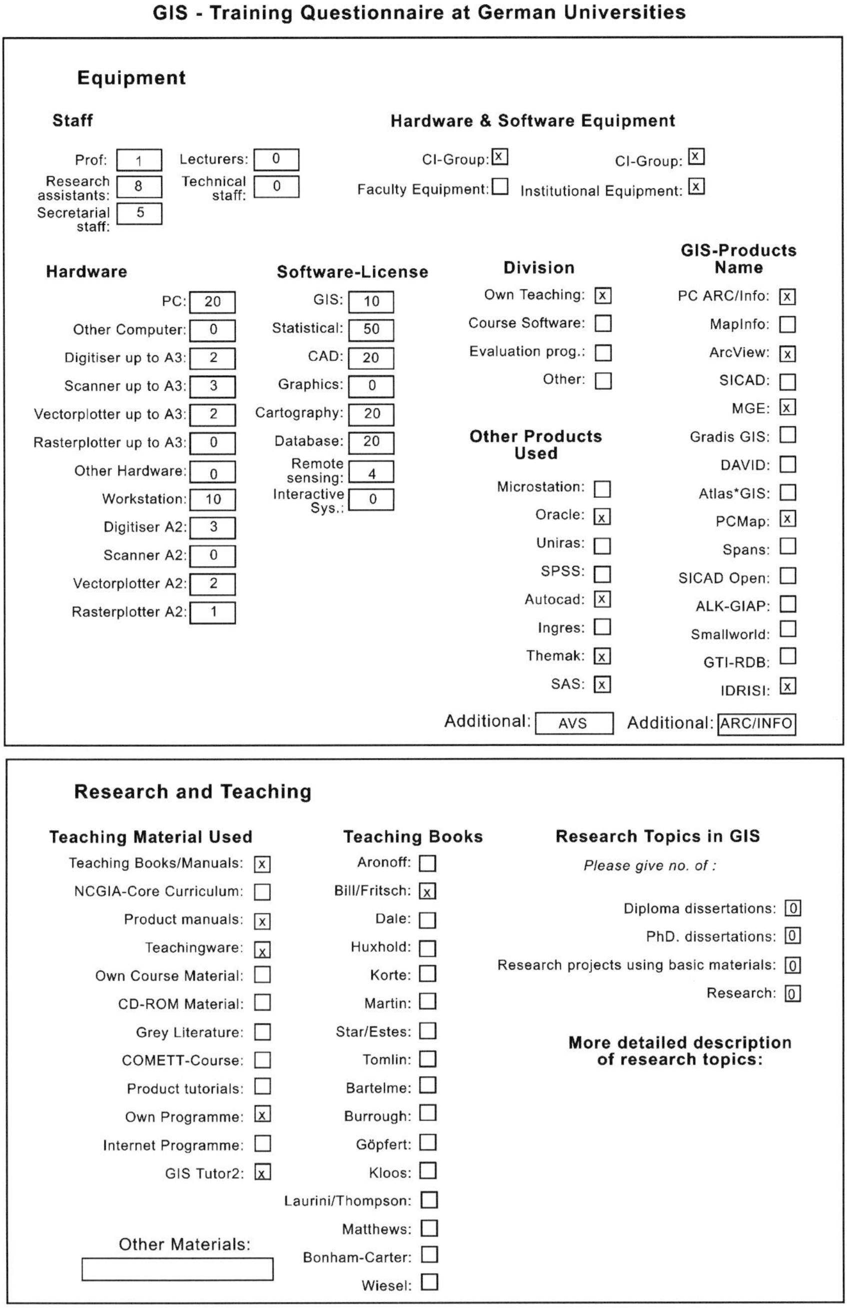

GIS - Training Questionnaire at German Universities

Equipment

Staff

Prof: 1   Lecturers: 0
Research assistants: 8   Technical staff: 0
Secretarial staff: 5

Hardware & Software Equipment

CI-Group: [x]   CI-Group: [x]
Faculty Equipment: [ ]   Institutional Equipment: [x]

Hardware

PC: 20
Other Computer: 0
Digitiser up to A3: 2
Scanner up to A3: 3
Vectorplotter up to A3: 2
Rasterplotter up to A3: 0
Other Hardware: 0
Workstation: 10
Digitiser A2: 3
Scanner A2: 0
Vectorplotter A2: 2
Rasterplotter A2: 1

Software-License

GIS: 10
Statistical: 50
CAD: 20
Graphics: 0
Cartography: 20
Database: 20
Remote sensing: 4
Interactive Sys.: 0

Division

Own Teaching: [x]
Course Software: [ ]
Evaluation prog.: [ ]
Other: [ ]

Other Products Used

Microstation: [ ]
Oracle: [x]
Uniras: [ ]
SPSS: [ ]
Autocad: [x]
Ingres: [ ]
Themak: [x]
SAS: [x]

Additional: AVS

GIS-Products Name

PC ARC/Info: [x]
MapInfo: [ ]
ArcView: [x]
SICAD: [ ]
MGE: [x]
Gradis GIS: [ ]
DAVID: [ ]
Atlas*GIS: [ ]
PCMap: [x]
Spans: [ ]
SICAD Open: [ ]
ALK-GIAP: [ ]
Smallworld: [ ]
GTI-RDB: [ ]
IDRISI: [x]

Additional: ARC/INFO

Research and Teaching

Teaching Material Used

Teaching Books/Manuals: [x]
NCGIA-Core Curriculum: [ ]
Product manuals: [x]
Teachingware: [x]
Own Course Material: [ ]
CD-ROM Material: [ ]
Grey Literature: [ ]
COMETT-Course: [ ]
Product tutorials: [ ]
Own Programme: [x]
Internet Programme: [ ]
GIS Tutor2: [x]

Other Materials:

Teaching Books

Aronoff: [ ]
Bill/Fritsch: [x]
Dale: [ ]
Huxhold: [ ]
Korte: [ ]
Martin: [ ]
Star/Estes: [ ]
Tomlin: [ ]
Bartelme: [ ]
Burrough: [ ]
Göpfert: [ ]
Kloos: [ ]
Laurini/Thompson: [ ]
Matthews: [ ]
Bonham-Carter: [ ]
Wiesel: [ ]

Research Topics in GIS

Please give no. of :

Diploma dissertations: [0]
PhD. dissertations: [0]
Research projects using basic materials: [0]
Research: [0]

More detailed description of research topics:

Figure 14.2 The education database interface

Environmental sciences   GIS is becoming more and more a compulsory subject, but not much time in total is spent on GIS except for a beginner's course or some seminars. More seminars (61 percent) are offered compared to lectures (29 percent) and a very small portion of exercises. Lectures are entitled GIS with specialisations in environmental information systems (EIS). GIS seminars include visualisation of environmental data, cartography (analogue, digital), remote sensing, terrain models and landscape information systems. A major exception is given in Vechta, where a post-graduate course on environmental monitoring is running with a major contribution on GIS (thirty-six hours) starting from the first half-year term.

Forestry   GIS in forestry is based more on application-driven lectures named GIS, environmental quality, PC-based map production and applied remote sensing.

Geodesy   Geodesy, together with geography, is handling the largest part of GIS-related topics during the courses of study: 54 percent lectures, and 28 percent exercises are given as mandatory subjects called GIS 1–3, GIS exercises, databases, and information systems (IS) or partially in photogrammetry, remote sensing, cartography and image processing. Especially at the FHS level some sites have recognised the chance to specialise in GIS. At FHS in Karlsruhe, Mainz and Stuttgart special courses of studies are offered in geoinformatics.

Geography   Geography is only offered at university level. GIS education shows more optional parts and is more oriented towards seminars and exercises. In geography together with geodesy, the largest part on GIS-related topics is part of the course studies. Lectures and seminars are entitled GIS 1–3, environmental information systems, and GIS in municipalities. Other lectures such as computer science, remote sensing, thematic cartography, digital image processing, computer cartography, landscape planning, landscape ecology, environmental planning, statistics and databases show a relevant part of GIS themes. The education is more application driven, running projects and seminars. Often GIS is set up as a course with a specific GIS product.

Landscape planning   GIS is an optional subject delivered in lectures (53 percent) and exercises (33 percent) on GIS and CAD, GIS 1–2, information systems, remote sensing, statistics, computer-based planning, graphical data processing, data processing in landscape planning. GIS is partly used in projects and again GIS product-specific seminars are usual.

Staff and equipment

Further evaluations of the questionnaire deal with the equipment available in institutions and relate to hardware, software and staff.

Academic staff

The number of academic staff is very restricted. On average there are two to three scientists per study course. This situation varies between the individual sites and courses. At FHS the scientific staff consists mainly of professors (52 percent) and their technical collaborators (21 percent) and supporting students (10 percent). Usually FHS do not have many scientific employees due to their target to offer a more practically oriented education. At universities GIS education and research is more attached to the scientific staff (51 percent) than to professors (16 percent). More student support (25 percent) is provided here, but fewer technicians (8 percent) are engaged in GIS education and research. Many institutions are dealing with GIS in a scientific way rather than from a lecturer's point of view.

Hardware

GIS platforms in education are mainly oriented towards PCs. Only one-third of the sites own workstations. Often only small digitisers, scanners and plotters are available. On average one large-format digitiser and plotter is available per study course. There is no major difference between FHS and university. The hardware belongs to the institute or department. Programmes from German research associations (so-called computer-investment-pools [CIP] and scientists' workplaces [WAP-pools]) have much improved hardware and software supplies for GIS education.

Software

GIS education and research strongly depends upon the availability of software packages. On average seven GIS licences can be used at the individual study course level. Further software packages include databases (Oracle®), statistical software (SPSS and SAS), cartographic, graphic (Uniras), and CAD software (AutoCAD and Microstation) as well as remote sensing products (ERDAS™ and PCI EASI/PACE®), which are similarly distributed at universities and FHS. Between the individual study courses major differences exist, however. Architecture, for example, uses more CAD packages whereas computer science is more oriented towards database software. Table 14.2 presents the ten basic GIS packages in use. The distributions between universities and FHS do not differ much.

Research topics

Various research topics are investigated at German institutions. At FHS most of the research work is done through diplomas because of the lack of scientific staff besides professors. At universities, diploma-based work, doctoral theses and funding from outside the university are of equal importance for research.

Table 14.2 The ten basic GIS software packages in use

GIS Products	Total number of sites
ArcView	43
PC Arc/Info	37
PCMap	22
IDRISI	20
Arc/Info	19
Atlas*GIS	15
SICAD and SICAD/open	12
MGE	11
SPANS®	8
ALK-GIAP	7
Smallworld	7

Only a small proportion of research is undertaken by the regular staff. The following research topics are related to the given groupings.

- Agricultural sciences: remote sensing, terrain relief, modelling of material flows in soil, water protection, precision farming, environmental damage and environmental protection
- Architecture, civil engineering and spatial planning: urban information systems, interfaces between CAD and GIS, environmental quality targets, GIS for citizens, GIS as the base of the planning and decision-making process
- Computer sciences: object-oriented GIS, 3-D and 4-D (space and time in GIS), query interfaces, fuzzy theory
- Economic sciences: GIS diffusion process in Germany
- Environmental sciences: environmental monitoring, environmental information systems, GIS in the planning process, databases for environmental samples
- Forestry: landscape information systems, hybrid GIS for forestry planning
- Geodesy: ALB, ALK and ATKIS (German land information systems), digital terrain models, remote sensing and GIS, digital cartography, raster-vector conversion and hybrid GIS, 3-D GIS, data quality, GIS and the Internet
- Geography: GIS and remote sensing, environmental modelling, location-allocation problems and spatial interaction, climatology, regional research, geology, geomorphology and terrain relief, ecology and ecosystems research
- Landscape planning: GIS in municipalities, GIS in landscape planning, environmental information systems

Conclusions

In Germany and especially in the new federal states, a lot of additional courses on GIS are up and running. Training courses on specific products are offered by software vendors; beginner's courses on GIS are given by universities, seminar centres and private education institutes. The remote study course of study, INTERGIS, is given in the German language by colleagues from the University of Salzburg (Austria, see Strobl and Blaschke, 1996). The full details of the research undertaken in this chapter will appear in Bill (1996).

Acknowledgements

The author would like to thank all colleagues for supplying the necessary information about their study courses.

References

Bill, R. (1992). 'On the Situation of GIS-Education at German Universities', in Proceedings of EGIS'92, Munich, pp. 846–55.

Bill, R. (1996). *GIS-Ausbildung an Deutschen Hochschulen – Ein Statusbericht*, Heft No. 1, Interne Berichte des Instituts fur Geodasie and Geoinformatik, Universität Rostock.

IFP (1992). *Vorträge zum Workshop Geoinformationssysteme in der Ausbildung Schriftenreihe*, Heft 16, Institut für Photogrammetrie an der Universität Stuttgart.

Strobl, J. and Blaschke, T. (1996). *GIS'96*, February 13–15, Wiesbaden, Institute for International Research.

Chapter 15

GIS in school education

An epilogue

David R. Green

Hopefully, the chapters in this book have succeeded in bringing some of the exciting world of GIS to both the teacher and the pupil. Together the contents of this book have offered a brief look at GIS. This has ranged from the simple practical manual approaches to the more complex approaches making use of software, based upon the practical experience of various individuals, many of whom are from a teaching or educational background with a new view on the use and value of spatial or geographical data and information.

Whilst developments in the UK education system have admittedly appeared to be somewhat slower and perhaps less clearly co-ordinated than in the US, some significant developments in software, teaching resources and experiences have nevertheless taken place and are a valuable addition to the overall teaching GIS resources available. However, it has to be said that the work carried out by both the NCGIA and the ESRI in the US, has also, in my view, offered the most promise to date, as far as GIS is concerned, to the school environment at both the elementary and secondary level. In fact the materials now available from these two sources are a very useful starting point for teachers and pupils in any country, in terms of the provision of a simple introduction to both GIS theory and applications as well as the provision of worked examples and datasets, some of which have now become more widely available over the Internet.

It is clear, however, that GIS software is not necessarily a prerequisite for the teaching of GIS in schools, and as the overhead example on the ESRI:K–12 web pages (http://www.esri.com/) prove, and as I pointed out in one of my chapters earlier in this volume (chapter 3), computers are not necessary at the outset of the GIS educational continuum. In many ways it is the basic theory that is important to convey to the pupil at first in order to provide a sound background of understanding that can then subsequently be taken further.

However, one must also remember that it is the computer technology that now defines GIS in the modern world, and whilst elements of the concepts of GIS can be conveyed practically both easily and effectively without the aid of a computer, the application of GIS theory and practical skills ultimately requires the use of computer technology, ideally using a Windows-based operating system. In fact, the transfer of theory into practice is vital and computer hardware and software really is necessary to achieve this in the classroom, if only to provide an illustration of how geographical data are handled, analysed, turned into information, displayed as map output, and incorporated or integrated with other IT in a decision-making environment.

As time has gone by, so hardware and software has become cheaper, more powerful and more sophisticated, and this has very rapidly enabled the possibilities of using software such as ArcView in schools, and more recently derivative products such as ArcExplorer. Running under MS Windows (3.1/95/98/2000/NT) these software packages have become much easier to use, and relatively more intuitive, allowing both teachers and pupils to explore spatial datasets and to undertake school projects based upon geographical data collection and analysis.

Whilst early examples of GIS applications for schools relied on simple software, and demonstration disks of applications from software suppliers, the ease of use and resources now to hand enables many pupils to make use of the 'real software' (the sort that is found in the workplace) as part of the educational curriculum, which in turn provides them with a basis upon which to pursue the use of the same and other software subsequently in further education or perhaps in the workplace.

Although it has unfortunately taken many years for many more people to recognise the potential of GIS, in part as a teaching tool, the delay has in many ways benefited everyone concerned. Microcomputer technology (hardware and software) has become much more sophisticated in the interim and much more widely available and affordable than ever before and this looks set to continue. A quick examination of some of the software we now use everyday on the desktop is testimony to this, and the hardware peripherals now available at relatively low cost such as scanners and digitisers, as well as digital cameras, have also acted as 'enablers' to allow GIS to spread into the classroom. The population at large is also generally far more computer literate than even a few years ago. Without such developments it would probably not have been possible to appreciate the importance of GIS in today's world.

In many ways we are also at one of the most exciting points ever in our history, and whatever our dislikes for today's world, the technology we now have at our fingertips offers children, our future generations, some of the most exciting possibilities ever in the context of information technology.

As David J. Maguire and Jack Dangermond once stated in a paper they wrote for me in the 1995 *AGI Sourcebook* (Maguire and Dangermond, 1994): 'The future is so bright I've got to wear shades', and subsequently David Rix and I wrote (Green and Rix, 1997): 'The future's so bright we all have to wear shades!' Today it would be appropriate to say that: 'The future is even brighter, we'll all have to wear shades!'

References

Green, D.R. and Rix, D. (1997). 'The Future', in *AGI Sourcebook for Geographic Information Systems 1997* (Eds D.R. Green, D. Rix and C. Corbin) pp. 55–7, AGI, London.

Maguire, D.J. and Dangermond, J. (1994). 'Future GIS Technology', in *AGI Sourcebook for Geographic Information Systems 1995* (Eds D.R. Green and D. Rix) pp. 113–20, AGI, London.

Glossary

The subject of geographic information has seen rapid development in recent years. It has introduced many new terms while at the same time producing new meaning to some existing terms in a particular context. Consequently, many publications include a glossary of terms for their own particular area of interest. The Standards Committee of the Association for Geographic Information published the first version of its GIS Dictionary in 1991. This dictionary attempts to produce a common set of definitions for general usage.

A new edition, published in 1993 and reproduced here, reflected the changes in the field of GIS that occurred since publication of the first edition. In that time, GIS became more a part of mainstream information technology (IT). This was reflected by the inclusion of more standard IT terms, especially those relating to databases and graphics, together with less emphasis on cartography.

The need to standardise terminology has been reflected in the work of both the International Standards Organisation (ISO) and the European Standards Committee (CEN), both of whom are producing sets of standard terminology. This internationalisation of GIS is reflected not only in the adoption of standard definitions where appropriate, but also in the removal from the dictionary of terms that were totally specific to the UK.

Where definitions are taken from a definitive source, that source is identified in brackets as a suffix. Full references are given to these sources in the list of abbreviations. Where no source is identified, then either the definition came from a non-definitive source, or it is a hybrid.

Comments received on the first edition were taken into account in the preparation of the 1993 edition. Any comments, errors or omissions should be notified to AGI, and will be taken into consideration in the preparation of the next edition.

The dictionary has been reprinted here as a glossary. The opportunity has been taken to correct some minor errors that occurred in the second edition published in October 1993.

Robert Walker, Editor
October 1993

Term	Definition
2.5D	A system in which the third dimension is constrained to a very simple relationship with the other two dimensions (e.g. where z is a single valued function of x and y).
Absolute Coordinate	One of the coordinates identifying the position of an addressable point with respect to the origin of a specified coordinate system. (ISO 2382-13)
Abstraction	A way of viewing a real world object. For example, a road may be a centre-line in one application and an area bounded by kerblines in another.
Accuracy	The closeness of results of observations, computations or estimates to the true values or the values accepted as being true. Accuracy relates to the exactness of the result, and is distinguished from precision which relates to the exactness of the operation by which the result is obtained.
Address	A means of referencing an object for the purposes of unique identification and location.
Adjacency	The sharing of a common side by two areas (polygons).
Aggregation	The grouping together of a selected set of like entities to form one entity. For example, grouping sets of adjacent areal units to form larger units, often as part of a spatial unit hierarchy (e.g. wards grouped into districts). Any attribute data is also grouped or is summarised to give statistics for the new spatial unit.
Algorithm	A finite, ordered set of well-defined rules for the solution of a problem. (ISO 2382-1)
Aliasing	Unwanted visual effects caused by insufficient sampling resolution or inadequate filtering to completely define the object, most commonly seen as a jagged edge along the object's boundary, or along a line. (ISO 2382-13)
Alphanumeric	Display of information in character format.
Analogue	In the context of remote sensing and mapping, the term refers to information in graphical or pictorial form as opposed to digital form. Generally, analogue refers to a quantity which is continuously variable, rather than one which varies only in discrete steps.
Annotation	The alphanumeric text or labels on a map, such as a street or place name.
ANSI	(American National Standards Institute) The US national standards body.

Application	A process that uses data from a system.
Applications package	A set of specialised programs and associated documentation, usually supplied by an outside agency or software house, to carry out a particular application (such as maintenance scheduling).
Arc	A line described by an ordered sequence of points and connections, between them defined by a mathematical function.
Archive	Accessible store for historical records and data.
Area	A bounded continuous two-dimensional object which may or may not include its boundary. Usually defined in terms of an external polygon or in terms of a set of grid cells. (AC)
ASCII	American Standard Code for Information Interchange. A standard binary coding system used to represent characters within a computer.
AM/FM	(automated mapping/facilities management) North American term for digital records, and applications thereof, particularly in the utilities.
Associated data	Alphanumeric information associated with a specific spatially referenced object.
Attribute	A trait, quality or property that is a characteristic of an entity. (EXPRESS)
Attribute class	A specified group of attributes. For example, those describing measure, serviceability, structure, or composition. (AC)
Attribute code	An alphanumeric identifier for an attribute. For example, 30-road type, Hg-concentration of mercury. (OS1)
Attribute value	A specific quality or quantity assigned to an attribute. For example, for the attribute 'type' the value 'steel'.
Automated cartography	The process of producing maps with the aid of computer-driven devices such as plotters and graphical displays. The term does not imply any information processing, beyond that required to support map production.
Automated digitising	Conversion of a map to digital form using a method which involves little or no operator intervention during the digitising stage, for example scanning. (OS1)
Automated feature recognition	The identification of map-based features using computer software incorporating pattern recognition techniques.
Automated mapping	See Automated cartography.

Base map	A set of topographic data displayed in map form, providing a frame of reference for users' data.
Basic spatial unit	(BSU) Fundamental areal unit having homogeneous properties in the context of any required theme such as administrative responsibility or ownership.
Benchmark	Subject to context:
	(i) A standard test devised to enable comparisons to be made between computer systems.
	(ii) Surveying term for a mark whose height relative to a datum is known.
Blind digitising	A method of manual digitising where the operator has no immediate graphic feedback to register progress. (OS1)
BLOB	(Binary large object) An area within a raster dataset that can be considered as a contiguous feature.
BLPU	(Basic land and property unit) The physical extent of a contiguous area of land under uniform property rights.
Buffer	A corridor of a specified width around a point, line or area.
Byte	A unit of computer storage of binary data usually comprising 8 bits, equivalent to a character. Hence Megabyte, one million bytes, and Gigabyte, one thousand million bytes.
CAD	(Computer-aided design) The design activities, including drafting and illustrating, in which information processing systems are used to carry out functions such as designing or improving a part or a product. (ISO 2382-24)
Cadastral survey	A survey based on the precise measurement and marking of land parcel boundaries.
Cadastre	The public register of the quality, value and ownership of the land of a country.
Cartography	The organisation and communication of geographically related information in either graphic or digital form. It can include all stages from data acquisition to presentation and use.
CCITT	(International Telegraph and Telephone Consultative Committee) An international body primarily addressing telecommunications standards.
Cell	The basic element of spatial information in the raster (grid square) description of spatial entities. (OS1)
CEN	(Comité Européen de Normalisation) The regional standards group for Europe. It is not a recognised

standards development organisation, and so cannot contribute directly to ISO. It functions broadly as a European equivalent of ISO and its key goal is to harmonise standards produced by the standards bodies of its member countries. Membership is open to EC and EFTA countries.

CENELEC (Comité Européen de Normalisation Electronique) This is the European equivalent of the IEC.

Centre line A line digitised along the centre of a linear feature. (OS2)

Centroid The position of the centre of gravity of an entity, often used to reference polygons.

CERCO (Comité Européen des Responsables de la Cartographie Officielle) The European Committee of Representatives of Official Cartography, under the auspices of the Council of Europe.

CGM (Computer Graphics Metafile) A standard (ISO 8632) file format specification for the storage and transfer of picture description information. (ISO 2382-13)

Chain A directed sequence of non-intersecting line segments and/or arcs with nodes at each end. (AC)

Chainage Distance measured along a link from a node.

Check plot A graphic output used to verify either the content or positional accuracy of digital data by direct super-imposition on the graphic original used to create the digital record. (AC)

Choropleth A thematic map shaded by value of a parameter.

Clipping The action of truncating data or an image by removing all the display elements that lie outside a boundary. (ISO 2383-13)

Compaction See data compression.

Complementary BSUs A family of BSUs designed around agreed definitions, the larger areas being the sum of a number of smaller areas etc.

Conceptual model A model that defines the types of entities or objects that are of immediate interest and the relationships between them.

Connectivity How the links and nodes in a network or polygons are joined to each other, and the degree to which it is applied (e.g. simply, multiply).

Contiguous Literally adjacent, touching. In the context of digital mapping, it implies a connected polygonal entity.

Continuous mapping A system of mapping in which the total extent of the mapped area is represented as a whole, without any spatial sub-divisions being apparent.

Contour	A set of points representing the same value of a given attribute and forming a line that may serve as a bounding of an area. (ISO 2382-13)
Control (mapping)	A system of points with established horizontal and vertical positions which are used as fixed references in positioning and relating map features. (AC)
Coordinated point	A point defined by a set of coordinates relative to a well-defined coordinate system.
Coordinates	Pairs of numbers expressing horizontal distances along orthogonal axes, or triplets of numbers measuring horizontal and vertical distances. (AC)
Currency	The level to which data is kept up to date.
Cursor	(i) A pointer appearing on a screen, under the control of a mouse or other pointing device.
	(ii) A hand-held device on a digitising table or tablet used for picking menus or accurately digitising graphic objects.
Data	A collection of facts, concepts or instructions in a formalised manner suitable for communication or processing by human beings or by automatic means.
Databank	A set of data related to a given subject and organised in such a way that it can be consulted by users. (ISO 2382-1)
Database	A collection of data organised according to a conceptual structure describing the characteristics of the data and the relationships among their corresponding entities, supporting applications areas. (ISO 2382-1)
Data capture	The encoding of data. In the context of digital mapping this includes digitising, direct recording by electronic survey instruments, and the encoding of text and attributes. (OS1)
Data classification	A description of the classes into which data is analysed or divided. (OS2)
Data compression	Methods of encoding which reduce the overall data volume. (See Run Length Encoding)
Data conversion	The transformation of data from paper records or digital form into a form suitable for loading into a GIS.
Data format	A specification that defines the order in which data is stored or a description of the way data is held in a file or record. (OS2)
Data item	A sequence of related characters which can be defined as the smallest logical unit of data that can be independently and meaningfully processed. For example, a single coordinate value. (OS1)

Data model	An abstraction of the real world which incorporates only those properties thought to be relevant to the application at hand. The data model would normally define specific groups of entities, and their attributes and the relationships between these entities. A data model is independent of a computer system and its associated data structures. A map is one example of an analogue data model. (OS1)
Data quality	Indications of the degree to which data satisfies stated or implied needs. This includes information about lineage, completeness, currency, logical consistency and accuracy of the data.
Dataset	An organised collection of data with a common theme.
Data structure	The logical arrangement of data as used by a system for data management; a representation of a data model in computer form. (OS1)
Data transfer	The movement of data from one system to another.
Datum	The fixed starting point of scale or a coordinate system.
DBMS	(DataBase Management System) A collection of software for organising the information in a database. Typically a DBMS contains routines for data input, verification, storage, retrieval and combination.
DEM	(Digital Elevation Model) Synonymous with Digital Terrain Model (DTM).
Derived map	A map produced from other maps rather than from an original survey.
Device coordinate	A coordinate specified by a device-dependent co-ordinate system. (ISO 2382-13)
Device space	The space defined by the complete set of addressable points of a display device. (ISO 2382-13)
Dialogue box	A pop-up window into which data may be entered. (ISO 2381-13)
DIGEST	The Digital Geographic Information Working Group (DGIWG) Exchange Standard. A NATO standard for the exchange of geographic data in digital form between defence agencies.
Digital chart of the world	A vector data set based on 1:1,000,000 scale air navigation charts, produced by the US Defence Mapping Agency.
Digital map data	The digital data required to represent a map. (OS1)
Digital mapping	The process of storing and displaying map data in computer form.

Digital records	Electronically stored representation of user data aligned with and stored as an overlay to a digital map base, together with associated information.
Digitising	The conversion of analogue maps and other sources to a computer readable form. (OS1) (This may be point digitising, where points are only recorded when a button is pressed on the puck, or stream digitising where points are recorded automatically at pre-set intervals of either distance or time.)
Digitising table	An electronic draughting table capable of recording the (x,y) coordinates of a point on a table in computer readable form.
Directed link	A link between two nodes with one direction specified. (AC)
Distributed data	Data held on several computers and accessible to users at other computers via communications networks.
Distributed system	A complex computer system where the workload is spread between two or more computers linked together by a communications network.
Domesday 2000	A project which aims to create a national computerised archive of property and land data in Britain.
Dpi	(dots per inch) A unit of measurement for the resolution of a scanning or printing device.
Dragging	Relocating display elements on a screen with a pointing device. This can be done by pressing and holding a pushbutton while moving the pointer on the screen. (ISO 2382-13)
DTM	(Digital Terrain Model) A digital representation of relief (ground surface). Usually a set of elevation values in correspondence with grid cells.
DX 90	A format for the supply of digital hydrographic data, developed by the International Hydrographic Organization (IHO). DX 90, together with the IHO's feature coding scheme (object catalogue) and a number of digitising conventions comprise the IHO transfer standard for digital hydrographic data.
DXF	(Digital exchange format) A format for transferring drawings between CAD systems, widely used as a de facto standard in the engineering and construction industries.
Echo	The immediate notification of the current values provided by an input unit to the user at the display console. (ISO 2382-13)
Edge	A line between two nodes, bounding one or more faces.

Edge matching	The process of ensuring that data along the adjacent edges of map sheets, or some other unit of storage, matches in both positional and attribute terms. (OS1)
EDI	(Electronic data interchange) The interchange of processable data between computers electronically.
EDIFACT	(Electronic data interchange for administration, commerce and transport) A set of syntax rules for the preparation of messages to be interchanged.
EDIGeO	A data transfer format strongly based on DIGEST, adopted by AFNOR as a French experimental standard Z-13-150.
Edit	The process of adding, deleting and changing data.
Emulator	A software package or program which imitates the functions of a hardware device or other software.
Encoding	The assignment of a unique code to each unit of information, such as encoding of English using the ASCII character set.
Entity	The general term for a real-world object or digital phenomenon.
Entity class	A specified group of entities. For example, hydrography, transportation. (AC)
Entity relationship model	A logical way of describing entities and their relationships.
Enumeration district	The basic area unit, containing approximately 150 households, used by the UK Office of Population Censuses and Surveys, for the planning and carrying out of population counts and surveys.
EUROGI	(European Umbrella Organisation for Geographic Information) An initiative to establish a single organisation covering the GIS community in Europe, made up from representatives of corporate organisations (associations).
EUROSTAT	The European Community statistical agency.
Expert system	A system that provides for solving problems in a particular application area by drawing inferences from a knowledge base acquired by human expertise. (ISO 2382-1)
Face	A surface bounded by a closed sequence of edges. Faces are contiguous and fill the spatial extent of the dataset and do not overlap.
Feature	A point or line used to represent one or more real-world objects. (OS2)
Feature code	An alphanumeric code which describes and/or classifies geographic features. (OS1)

Feature serial number — A unique code to identify an individual feature.

Fenceline — Boundary of a parcel of land physically represented on the ground by a surveyable object such as a wall or a fence.

Field — A set of one or more characters comprising a unit of information.

File — A named set of records stored or processed as a unit. (ISO 2382-1)

Format — A language construct that specifies the representation, in character form, of objects on a file.

Generalisation — Simplification of map information, so that information remains clear and uncluttered when map-scale is reduced. Usually involves a reduction in detail, a resampling to larger spacing, or a reduction in the number of points in a line. (OS2)

Geocode — A code which represents the spatial characteristics of an entity. For example, a coordinate point or a postcode. (OS1)

Geodesy — The science of measuring the Earth.

Geodetic datum — The definition of a particular spheroid and its position and orientation relative to the geoid.

Geographic information — Information about objects or phenomena that are associated with a location relative to the surface of the Earth. A special case of spatial information.

Geoid — An imaginary shape for the Earth defined by mean sea level and its imagined continuation under the continents at the same level of gravitational potential.

Geometry — The shape of the represented entity or entities, in terms of its stored coordinates. (OS2)

GIS — (Geographic information system) A system for capturing, storing, checking, integrating, manipulating, analysing and displaying data which are spatially referenced to the Earth. (OS1)

GKS — (Graphics Kernal System) A standard specification (ISO 7942) for a set of functions for computer graphics programming, and a functional interface between an applications program and the graphical input and output devices. (ISO 2382-13)

GPS — (Global positioning system) A constellation of US satellites which enables users with appropriate receivers to fix their position on or above the surface of the Earth to varying degrees of accuracy depending on the receivers and techniques used.

Graphic primitive	A basic graphic element that can be used to construct a display image (e.g. point, line segment). (ISO 2382-13)
Graphical user interface	A user interface which makes use of graphical objects such as icons, for selecting options, and usually has a windowing capability, enabling multiple window displays on the same screen.
Graphics tablet	A special flat surface with a mechanism for indicating positions thereon, normally used as a locator. (ISO 2382-13)
Graticule	The depiction of the lines of latitude and longitude on a map. The lines will not be orthogonal or even, in general, straight.
Grey scale	A range of intensities between black and white. (ISO 2382-13)
Grid	An orthogonal set of lines depicting a plane co-ordinate system.
Grid cell	A two-dimensional object that represents an element of a regular or nearly regular tessellation of a surface. (AC)
Grid reference	The position of a point on a map expressed in terms of grid co-ordinates.
Grid squares	A regular array of square cells referenced to a grid, used as a basis for holding spatially referenced information.
Hard copy	A print or plot of output data on paper or some other tangible medium. (OS1)
Hardware	All or part of the physical components of an information processing system. (ISO 2382-1)
Hidden line	A line or segment of a line which can be masked in a view of a three-dimensional object. (ISO 2382-13)
Highlighting	Emphasising a display element by modifying its visual attributes. (ISO 2382-13)
Hotspot	The (x,y) position that corresponds to the coordinates reported for a pointer. (ISO 2382-13)
Icon	A graphic symbol, displayed on a screen, that the user can point to with a device such as a mouse, in order to select a particular function or software application. (ISO 2382-13)
IEC	(International Electrotechnical Committee) This has the same status as ISO, but focuses on electrical and electrotechnical issues, especially electricity measurement, testing, use and safety.

IEEE	(The Institute of Electrical and Electronics Engineering Inc.) A major international professional body and an accredited standards setting organisation.
IGES	(International graphics exchange system) An ANSI standard for the exchange in digital form of CAD drawings.
Image	A raster representation of a graphic product (scanned map, photograph, drawing etc.) or a remotely sensed surface that consists of one or more spectral bands.
Image processing	The use of a data processing system to create, scan, analyse, enhance, interpret or display images. (ISO 2382-1)
Information	Intelligence resulting from the assembly, analysis or summary of data into a meaningful form.
Integration	The combining of data of different types from different sources and systems to provide new information.
Interactive digitising	A method of digitising in which dialogue (interaction) takes place between the operator and the computer. (OS1)
Interface	The junction or linking together of two or more systems.
Interior area	An area not including its boundary. (AC)
Interoperability	The capability to communicate, execute programs, or transfer data among various functional units (items of hardware or software or both, capable of accomplishing a specified purpose) in a manner that requires the user to have little or no knowledge of the unique characteristics of those units. (ISO 2382-1)
Invisible line	A line which is sometimes used to make a logical connection between two parts of a feature or between two different features but which is not displayed. (OS1)
IRDS	(Information resource dictionary system) An international standard (IS) which includes data dictionaries and software development aspects of building database management systems.
Island	An area delimited by a contour and surrounded by a fill pattern. (ISO 2382-13)
ISO	(International Organization for Standardization) The main de jure international standards setting body.
Isoline	A line joining points of equal value of a parameter.
Land parcel	An area of land, usually with some implication for land ownership or land use. (OS1)

Land terrier	A document system comprising a set of marked maps and ledgers containing textual information to record land and property.
Latitude	The angular distance between the normal at a point on an ellipsoid and the plane of the equator of the ellipsoid.
Layer	A usable subdivision of a dataset, generally containing objects of certain classes. (See also, level.)
Level	A term synonymous with Layer.
Line	A series of connected coordinated points forming a simple feature with homogeneous attribution.
Line segment	A line described by two sets of coordinates and the shortest connection between them.
Lineage/origin	The ancestry of a dataset describing its origin and the processes by which it was derived from that origin. Lineage is synonymous with provenance, but is more than just the original source or author.
Link	A line without logical intermediate intersections.
Link and node structure	A data structure in which links and nodes are stored with cross-referencing. (OS1)
LIS	(Land information system) A system for capturing, storing, checking, integrating, manipulating, analysing and displaying data about land and its use, ownership, development, etc. (OS1)
Locational reference	The means by which information can be related to a specific spatial position or location.
Locator	An input unit, such as a graphics tablet or any pointing device, that provides data to generate coordinates of a position. (ISO 2382-13)
Longitude	The angular distance between a meridian plane through a point on an ellipsoid, and an arbitrarily defined meridian (usually the Greenwich meridian).
Manual digitising	A method of digitising by an operator moving a cursor over a map on a digitising table. (OS1)
Map	A graphic representation of features of the Earth's surface or other geographically distributed phenomena. Examples are topographic maps, road maps, weather maps.
Map projection	A systematic portrayal of geographically distributed features from the (curved) surface of the Earth onto a plane.
MEGRIN	(Multi-purpose European Ground-Related Information Network) A consortium of European countries developing joint topographic data such as national and administrative boundaries.

Menu	A list of options displayed by a data processing system, from which the user can select an action to be initiated. (ISO 2382-1)
Menu bar	An area along one edge of a window used to display names or icons for menus. (ISO 2382-13)
Mereing	The definition of a boundary in relation to topographic features on the ground at the time of survey (e.g. 'one metre from the road edge'). If at a later date the ground detail changes or disappears, then that section is not re-mered, but the mereing changes to 'defaced'.
Meta data	Information about data. Examples are data quality information or feature classification information. (OS2)
Meta model	A model that defines the components (concepts) needed to define conceptual models (application models). The Meta Model also defines the relationships between the components.
Model	A formal description of the real world or part of it. (EXPRESS)
Mouse	A device used for pointing and selecting areas of a VDU or graphics screen, moved over another surface.
Multimedia	A combination of a variety of user interfaces and communication elements such as still and moving pictures, sound, graphics and text.
National Grid	The metric grid on a Transverse Mercator Projection used by the Ordnance Survey to provide an unambiguous spatial reference in Great Britain for any place or entity whatever on the map scale. (OS1)
Network	An arrangement of nodes and interconnecting links.
Node	The start or end of a link or line. A node may be shared by several lines. (OS1)
NTF	(National Transfer Format) A British Standard (BS 7567) for the transfer of geographic data, administered by AGI.
Object	(i) A discrete element of the real world, instances of which occur in a dataset.
	(ii) A collection of entities which form a higher level entity. For example, an object could be built from a collection of links to form a polygon. (OS2)
Object class	A logical classification of objects (e.g. buildings, roads). Classes may contain sub-classes, or may be sub-classes of other classes (e.g. motorways are a sub-class of roads, which are a sub-class of transport systems).

Object-oriented programming	The writing of computer programs using object-oriented techniques and languages. These employ a data-centred approach to programming, based on objects and object classes. Data is 'encapsulated' with operations, and commands are executed using message passing.
OEEPE	(Organisation Européene d'Etudes en Photogrammetrie Experimentale) A voluntary group of individuals and organisations that carry out research into photogrammetry and related subjects.
Open system	An information processing system that complies with the requirements of open systems interconnection standards in communication with other such systems.
Origin	The reference point (0,0) from which coordinates are measured. (OS2)
OSI	(Open systems interconnection) This defines the accepted international standards (IS 7498-1984) by which open systems should communicate with each other. It takes the form of a seven-layer model of a network architecture, with each layer performing a different function.
OSTF	(Ordnance Survey transfer format) A UK data transfer format previously used by Ordnance Survey for the supply of digital data to customers.
Overlay	A set of graphical data that can be superimposed on another set of graphical data through registration to a common coordinate system.
Overshoot	The projection of a line feature beyond the true point of intersection with another line feature. (OS2)
Panning	Progressively translating the display elements to give the visual impression of lateral movement of the image. (ISO 2382-13)
Parcel	See Land parcel.
Pecked line	A line drawn as a series of dashes. (OS1)
PHIGS	(Programmer hierarchical interactive graphics system) A standard (ISO 9592) set of graphics support functions to control the definition, modification, storage, and display of hierarchical graphics data. (ISO 2382-13)
Photogrammetry	The measurement of photographic images for the purpose of extracting useful information, particularly for the creation of accurate maps.
Pixel	The smallest element of a display surface that can be independently assigned attributes such as colour and intensity. (ISO 2382-13)

Point	A zero-dimensional abstraction of an object, with location specified by a set of coordinates.
Pointer	A symbol displayed on a screen, that a user can move with a pointing device such as a mouse, to select items. (ISO 2382-13)
Polygon	An area bounded by a closed line. (OS2)
Polyline	A line made up of a sequence of line segments.
Polymorphism	The ability to use the same consistent methods in different situations.
Pop-up window	A window that appears rapidly on the display surface in response to some action. (ISO 2382-13)
Portability	The capability of a program to be executed on various types of data processing systems without converting the program to a different language and with little or no modification. (ISO 2382-1).
Positional accuracy	The degree to which a position is measured or depicted, relative to its correct value established by a more accurate process.
Postcode	A coding system for referencing all properties in the UK which have a postal address. The country is divided into 120 areas, each area is divided into districts, each district into sectors, each sector into units. A unit postcode applies to a group of adjacent addresses (approximately 15 in number) of neighbouring properties and does not define an area.
Precision	The exactness with which a value is expressed, whether the value be right or wrong. (OS1)
Pre-development mapping	A concept in which one party prepares a definitive digital map of a proposed development which is then used by all interested parties for detailed planning and recording purposes until the as-built has been surveyed.
Premcode	A set of alphanumeric characters which, when added to the unit postcode, serves to identify a specific premise or address in the UK.
Program	A logically connected set of instructions which tell a computer to perform a sequence of tasks.
Projection	See Map projection.
Property	Land, building or estate which is owned and to which legal title exists.
Protocol	The method by which components of a system communicate with each other.
Puck	A pointing device that must be positioned manually on the pad of a graphics tablet in order to register input points when tracing images. (ISO 2382-13)

Pull-down menu	A menu that appears below the menu bar when the user selects a name or icon from the menu bar. (ISO 2382-13)
Quadtree	The expression of a two-dimensional object as a tree structure of quadrants which are formed by recursively subdividing each nonhomogeneous quadrant until all quadrants are homogeneous with respect to a selected property, or until a predetermined cutoff depth is reached. (ISO 2382-13)
Raster data	Spatial data expressed as a matrix of cells or 'pixels', with spatial position implicit in the ordering of the pixels. (OS2)
Raster scan	A technique for generating or recording the elements of a display image by means of a line-by-line sweep across the entire display space; for example, the generation of a picture of a television screen. (ISO 2382-13)
Raster to vector conversion	The process of converting an image made up of cells into one described by lines and polygons.
Ray tracing	A technique for determining, by tracing imaginary rays of light from the viewer's eye to the objects in a scene, the parts of the scene that should be displayed in the resulting image at any given point in time. (ISO 2382-13)
RDBMS	(Relational DataBase Management System) A database management system that supports the relational model.
Real-time system	A system that is able to receive continuously changing data from external sources and to process that data sufficiently rapidly to be capable of influencing the sources of data (e.g. monitoring reservoir levels).
Record	A set of related data fields grouped for processing.
Relative accuracy	The measure of the internal consistency of the positional measurements in a dataset. For many local area purposes, for example records of utility plant, relative accuracy is more important than absolute accuracy. In this case, accurate measurement of offsets from fixed points is required rather than knowledge of the true position in space.
Relative co-ordinate	One of the coordinates identifying the position of an addressable point with respect to another addressable point. (ISO 2382-13)
Remote sensing	The technique of obtaining data about the environment and the surface of the Earth from a distance, for example, from aircraft or satellites. (OS1)

Rendering	The conversion of the geometry, colouring, texturing, lighting and other characteristics of a scene into a display image. (ISO 2382-13)
Repeatability	The ability of a device to perform the same action consistently or to provide the same data given identical conditions. Given identical inputs, the limits within which the output will fall with a given statistical confidence. (OS1)
Resolution	A measure of the ability to detect quantities. High resolution implies a high degree of discrimination but has no implication as to accuracy. (OS1)
Ring	A sequence of non-intersecting chains, strings, links, or arcs with closure. (It represents a closed boundary, but not the interior area inside the closed boundary.) (AC)
RLE	(Run Length Encoding) The process of encoding a digital data stream which defines that stream in terms of the number of successive digital data elements that have the same value. (ISO 2382-13)
Rubber banding	The result of moving a point or an object in a manner that preserves interconnectivity with other points or objects through stretching, shrinking or reorienting their interconnecting lines. (ISO 2382-13)
Scale	The ratio or fraction between the distance on a map, chart or photograph and the corresponding distance in the real world.
Scanning	A method of data capture whereby an image or map is converted into digital raster form by systematic line-by-line sampling.
Schema	A collection of items forming part or all of a model. (EXPRESS)
SDTS	(Spatial Data Transfer Standard) The US Federal Information Processing Standard for the transfer of spatial data (FIPS Publication 173).
Seed	A point within an area that can be used to carry the attributes of the whole area, e.g. ownership, address, land use type. (OS1)
Semi-automatic digitising	A method of digitising from a map in which the majority of the line-following is controlled by a machine, but which requires an operator to be on hand constantly to assist the machine to identify features and resolve anomalies. (OS1)
SIF	(Intergraph's Standard Interchange Format) Proprietary format primarily used for the transfer of CAD drawings.

Sliver polygon	A small area formed when two polygons which have been overlaid do not abut exactly, but overlap along one or more edges.
Snapping	An operation or process whereby the computer will pick a nearby point or closest point on a nearby line.
Software	All or part of the programs, procedures, rules and their associated documentation of an information processing system. (ISO 2382-1)
Software engineering	The systematic application of scientific and technological knowledge, methods and experience to the design, implementation, testing and documentation of software to optimise its production, support and quality. (ISO 2382-1)
Software package	A fully documented program or set of programs, designed to perform a particular task (e.g. a word processor).
Spaghetti	Vector data composed of line segments which are not topologically structured or organised into objects and which may not even be geometrically tidy.
Spatial analysis	Analytical techniques associated with the study of locations of geographic phenomena together with their spatial dimensions.
Spatial information	Information which includes a reference to a two- or three-dimensional position in space as one of its attributes.
Spatial reference	Co-ordinate, textual description or codified name by which information can be related to a specific position or location on the Earth's surface.
Spline	A smooth curve fitted mathematically to a sequence of points.
SQL	(Structured query language) An ISO standard (IS 9075) interface to relational database products. It is used to define and access databases and to manipulate the data stored in them.
STEP	A transfer format for graphics data being developed by ISO (TC184/SC4) to replace IGES.
String	A sequence of line segments or text items. (A string does not have nodes, node identifiers, or left and right identifiers and may intersect itself or other strings.) (AC)
Surveying	The measurement and recording of geographically distributed information. Particular types of surveying are topographic, cadastral and geological.
Symbol	A graphic representation of a concept that has meaning in a specific context. (ISO 2383-1)

Systems development methodology	An integrated set of techniques and methods for effective and efficient planning, analysis, design, construction, implementation and support of computer systems.
Tablet	See Graphics tablet.
Tessellation	The subdivision of a two-dimensional plane (or three-dimensional volume) into disjoint congruent polygonal tiles (polyhedral blocks). (OS1)
Tesseral	A gridded representation of the plane surface into disjoint polygons. These polygons are normally either square (raster), triangular (TIN), or hexagonal. These models can be built into hierarchical structures, and have a range of algorithms available to navigate through them.
Thematic map	A map depicting one or more specific themes. Examples are land classification, population density, rainfall etc.
Thiesen polygon	A polygon bounding the region closer to a point than to any adjacent point.
TIGER	(Topologically integrated geocoding and referencing) A data format developed by the US Bureau of Census for the 1990 US census.
Tile	A logical rectangular set of data used to subdivide digital map data into manageable units. (OS2)
TIN	(Triangulated irregular network) A form of the tesseral model based on triangles. The vertices of the triangles form irregularly spaced nodes. Unlike the grid, the TIN allows dense information in complex areas, and sparse information in simpler or more homogeneous areas.
Tint	A stipple dot pattern used to create subdued colour infill within a defined area. (OS2)
Topographical database	A database in which data relating to the physical features and boundaries on the Earth's surface is held. (OS1)
Topologically structured data	Data structured such that relations and characteristics referred to as topology can be expressed, including concepts such as connectivity, adjacency and containment.
Topology	The relative location of geographic phenomena independent of their exact position. In digital data, topological relationships such as connectivity and relative position are usually expressed as relationships between nodes, links and polygons. (OS2)

Transfer format	The format used to transfer data between computer systems. In general usage this can refer not only to the organisation of data, but also to the associated information, such as attribute codes which are required in order to successfully complete the transfer. (OS1)
Transfer medium	The physical medium on which digital data is transferred from one computer system to another. For example, magnetic tape. (OS1)
Transformation	A computational process of converting a position from one coordinate system to another. (AC)
Undershoot	A line feature which is short of its true intersection with another line feature. (OS2)
Universe of discourse	That aspect of the real world under consideration.
Update	The process of adding to and revising existing information to take account of change. (OS1)
UPRN	(Unique Property Reference Number) A short-coded address or number which uniquely identifies a parcel of land.
Vector data	Positional data in the form of coordinates of the ends of line segments, points, text position etc. (OS1)
Window	A part of a display image with defined boundaries in which information is displayed. (ISO 2382-13)
Wireframe representation	A representation of an object, composed entirely of lines as though constructed of wire. (The lines may represent edges or surface contours in the display including those that may be hidden in the view of a real object.) (ISO 2382-13)
Workstation	An individual interactive computer with at least a processor, screen and keyboard, that is more powerful than a PC and typically operates under a multitasking operating system.
WYSIWYG	(What-You-See-Is-What-You-Get) A capability to display information on a screen, exactly as it will be printed or plotted on an output device.
Zooming	Progressively scaling the entire display image to give the visual impression of movement of display elements towards or away from the observer. (ISO 2382-13)

Sources (and abbreviations)

AC	*American Cartographer*, Vol. 15 No. 1, Jan. 1988.
EXPRESS	ISO CD 10303-11, *Product Data Representation and Exchange – Part 11: The EXPRESS Language Reference Manual*, 1991.

ISO 2382-1	*Information Technology – Vocabulary – Part 1: Fundamental Terms*, 1992.
ISO 2382-13	*Information Technology – Vocabulary – Part 13: Computer Graphics*, 1992.
ISO 2382-24	*Information Technology – Vocabulary – Part 24: Computer Integrated Manufacture*, 1992.
OS1	Working Party to produce National Standards for the transfer of digital data. Glossary of Terms, January 1987.
OS2	Ordnance Survey. Sample object data specification. March 1993.

Reading list

This is not intended to be a comprehensive reading list of books but is intended to cover a broad range of the literature currently available pitched at a level suitable for background reading. Some books available in publishers catalogues have been excluded on the grounds that they are a little too specialist for the intended readership of this volume. Whilst most of the books listed below go well beyond the school curriculum they do provide useful information through selective reading, valuable illustrations, and in addition further contacts, insight and direction as far as the higher education curriculum is concerned. The reading list will be updated on the website (www.abdn.ac.uk/gis-school/).

GIS

Aronoff, S. (1989). *Geographic Information Systems: A Management Perspective*, WDL Publications, Ottawa.

Bernhardsen, T. (1992). *Geographic Information Systems*, VIAK IT, Norway.

Berry, J.K. (1993). *Beyond Mapping: Concepts, Algorithms and Issues in GIS*, GIS World Inc., Fort Collins, CO.

Berry, J.K. (1995). *Spatial Reasoning for Effective GIS*, GIS World Inc., Fort Collins, CO.

Birkin, M., Clarke, G., Clarke, M. and Wilson, A. (1996). *Intelligent GIS: Location Decisions and Strategic Planning*, Longman, London.

Burrough, P.A. (1986). *Principles of Geographical Information Systems for Land Resources Assessment*, Clarendon Press, Oxford.

Campbell, H.J. and Masser, I. (1995). *GIS and Organizations: How Effective are GIS in Practice?* Taylor & Francis, London.

Carver, S. (Ed.) (1998). *Innovations in GIS 5*, Taylor & Francis, London.

Cassettari, S. (1993). *Introduction to Integrated Geo-Information Management*, Chapman & Hall, London.

Chrisman, N.R. (1997). *Exploring Geographic Information Systems*, John Wiley & Sons, New York.

Clarke, K.C. (1990). *Analytical and Computer Cartography*, Prentice-Hall, Englewood Cliffs, New Jersey.

Davis, B. (1996). *GIS: A Visual Approach*, OnWord Press, Santa Fe.

DeMers, M.N. (1997). *Fundamentals of Geographic Information Systems*, John Wiley & Sons, New York.

ESRI (1990). *Arc/Info Maps*. Environmental Systems Research Institute, Redlands, CA.

ESRI (1998). *ESRI Map Book* – Volume Thirteen. Environmental Systems Research Institute, Redlands, CA.

Fisher, P. (Ed.) (1995). *Innovations in GIS 2*, Taylor & Francis, London.

Fotheringham, A.S. and Rogerson, P. (Eds) (1994). *Spatial Analysis and GIS*, Taylor & Francis, London.

Goodchild, M.F. and Gopal, S. (Eds) (1989). *The Accuracy of Spatial Databases*, Taylor & Francis, London.

Green, D.R., Rix, D. and Corbin, C. (1997). *The AGI Sourcebook for Geographic Information Systems 1997*, AGI, London.

Grimshaw, D.J. (1994). *Bringing Geographical Information Systems into Business*, Longman, London.

Haines-Young, R., Green, D.R. and Cousins, S.H. (Eds) (1993). *Landscape Ecology and GIS*, Taylor & Francis, London.

Hearnshaw, H.M. and Unwin, D.J. (Eds) (1994). *Visualization in Geographical Information Systems*, John Wiley & Sons, New York.

Heit, M. and Shortreid, A. (1991). *GIS Applications in Natural Resources*, GIS World Inc., Fort Collins, CO, 381p.

Hohl, P. and Mayo, B. (1997). *ArcView GIS Exercise Book*, OnWord Press, Santa Fe.

Kemp, Z. (1997). *Innovations in GIS 4*, Taylor & Francis, London.

Kennedy, M. (1996). *The Global Positioning System and GIS: An Introduction*, Ann Arbor Press, Ann Arbor, MI.

Korte, G.P. (1997). *The GIS Book*, OnWord Press, Santa Fe.

Langran, G. (1992). *Time in Geographical Information Systems*, Taylor & Francis, London.

Laserna, R. and Landis, J. (1989). *Desktop Mapping for Planning and Strategic Decision-Making*. Strategic Mapping Inc., San Jose, CA.

Laurini, R. and Thompson, D. (1992). *Fundamentals of Spatial Information Systems*, Academic Press, London.

Longley, P. and Clarke, G. (Eds) (1995). *GIS for Business and Service Planning*, GeoInformation International, Cambridge, UK.

Maguire, D.J. (1989). *Computers in Geography*, Longman, London.

Maguire, D.J., Goodchild, M.F. and Rhind, D.W. (Eds) (1991). *Geographical Information Systems: Principles and Applications*, Longman, London.

Martin, D. (1996). *Geographical Information Systems and their Socio-economic Applications*, Routledge, London.

Mather, P.M. (1991). *Computer Applications in Geography*, John Wiley & Sons, London.

Mather, P.M. (Ed.) (1993). *Geographical Information Handling – Research and Applications*, John Wiley & Sons, Chichester.

McDonnell, R. and Kemp, K.K. (1995). *International GIS Dictionary*, GeoInformation International, London.

Plewe, B. (1997). *GIS Online: Information Retrieval, Mapping and the Internet*, OnWord Press, Santa Fe.

Ordnance Survey. *Ordnance Survey – Teacher Resource File – Port Talbot*. Ordnance Survey Educational Publishing.

Parker, D. (1996). *Innovations in GIS 3*, Taylor & Francis, London.

Peuquet, D.J. and Marble, D.F. (1990*). Introductory Readings in Geographical Information Systems*, Taylor & Francis, London.

Pickles, J. (Ed.) (1995). *Ground Truth: The Social Implications of Geographic Information Systems*, Guildford Press, New York.

Price, M.F. and Heywood, D.I. (Eds) (1994). *Mountain Environments and Geographic Information Systems*, Taylor & Francis, London.

Raper, J.F., Rhind, D.W. and Shepherd, J.W. (1992). *Postcodes: The New Geography*, Longman, London.

Reeve, D.E. and Petch, J.R. (1998). *GIS Organizations and People: A Socio-Technical Approach*, Taylor & Francis, London.

Rhind, D. (1998). *Framework for the World*, John Wiley & Sons, UK.

Taylor, D.R.F. (1991). *Geographic Information Systems: The Microcomputer and Modern Cartography*, Pergamon Press, Oxford.

Tomlin, C.D. (1990). *Geographic Information Systems and Cartographic Modelling*, Prentice-Hall: Englewood Cliffs, NJ.

Tomlinson, R.F., Calkins, H.W. and Marble, D.R. (1976). *Computer Handling of Geographic Data*, UNESCO Press, Paris.

Unwin, D. (1981). *Introductory Spatial Analysis*, Methuen, London.

Verbyla, D.L. and Kang-Tsung, C. (1997). *Processing Digital Images*, OnWord Press, Santa Fe.

Whitener, A., and Creath, B. (1996). *Mapping with Microsoft Office*, OnWord Press, Santa Fe.

Worboys, M.F. (1995). *GIS: A Computing Perspective*, Taylor & Francis, London.

Zeiler, M. (1994). *Inside Arc/Info*, OnWord Press, Santa Fe.

GIS, education and training

Coggins, P.C. (1990). 'Horses for Courses', Proceedings of Mapping Awareness 1990 Conference.

Goodbrand, C. (1991). 'Educational and Geographical Information Systems', *GIS Applications in Natural Resources,* GIS World Inc., Fort Collins, CO.

Keller, C. P. (1991). 'Issues to Consider when Developing or Selecting a GIS Curriculum', *GIS Applications in Natural Resources*, GIS World Inc., Fort Collins, CO.

King, G.Q. (1991). 'Geography and GIS Technology', *Journal of Geography*, pp. 66–72.

Miller, W.R. (1992). 'Creating the New Geographer', *ARC NEWS*, Education News section, Winter, p. 6.

Sullivan, S.A. and Miller, C.R. (1991). 'GIS Training and Education: The Need for a New Approach', *GIS Applications in Natural Resources*, GIS World Inc., Fort Collins, CO.

GIS defined

Cowen, D.J. (1990). 'GIS versus CAD versus DBMS: What are the Differences', in Peuquet, D.J. and Marble, D. (Eds) (1990), *Introductory Readings in GIS,* pp. 52–62, Taylor & Francis, London.

Dangermond, J. (1990). 'A Classification of Software Components Commonly Used in GIS', in Peuquet, D.J. and Marble, D. (Eds) (1990), *Introductory Readings in GIS*, pp. 30–51, Taylor & Francis, London.

ESRI (1991). *Understanding GIS*, pp. 1-1–1-10, Environmental Systems Research Institute, Redlands, CA.

Marble, D. (1990). 'Geographic Information Systems: An Overview', in Peuquet, D.J. and Marble, D. (Eds) (1990), *Introductory Readings in GIS*, pp. 8–17, Taylor & Francis, London.

Milne, P.H. (1991). 'CAD: an Input to GIS', *Mapping Awareness*, Vol. 5(7), September, pp. 32–34.

Newell, R.G. and Theriault, D.G. (1990). 'Is GIS Just a Combination of CAD and DBMS?', *Mapping Awareness*, Vol. 4(3), pp. 42–5.

Peuquet, D.J. and Marble, D. (Eds) (1990). *Introductory Readings in Geographic Information Systems*, Taylor & Francis, London, 371p.

Peuquet, D.J. and Marble, D. (1990). 'What is a Geographic Information System', in Peuquet, D.J. and Marble, D. (Eds) (1990), *Introductory Readings in GIS*, pp. 5–7, Taylor & Francis, London.

Ripple, W.J. (Ed.) (1989). *Fundamentals of Geographic Information Systems: A Compendium*, pp. 3–7, American Society for Photogrammetry and Remote Sensing/ACSM, Bethesda, MD.

Tomlinson, R.F. (1990). 'Geographic Information Systems – A New Frontier', in Peuquet, D.J. and Marble, D. (Eds) (1990), *Introductory Readings in GIS*, pp. 18–29, Taylor & Francis, London.

Walsh, S. (1988). 'Geographical Information Systems: An Instructional Tool for Earth Science Educators', *Journal of Geography*, Vol. 12(5), pp. 17–25.

Remote sensing

Avery, T.E., and Berlin, G.L. (1992). *Interpretation of Aerial Photographs*, Prentice-Hall, Englewood Cliffs, NJ.

Barrett, W.C. and Curtis, L.F. (1982). *Introduction to Environmental Remote Sensing*, Chapman & Hall, London.

Campbell, J.B. (1996). *Introduction to Remote Sensing*, Taylor & Francis, London.

Cracknell, A.P. and Hayes, L.W.B. (1990). *Introduction to Remote Sensing*, Taylor & Francis, London.

Curran, P.J. (1985). *Principles of Remote Sensing*, Longman, London.

Harris, R. (1987). *Satellite Remote Sensing: An Introduction*, Routledge, London.

Jensen, J.R. (1996). *Introductory Digital Image Processing – A Remote Sensing Perspective*, Prentice-Hall, Englewood Cliffs, NJ.

Lillesand, T. and Kiefer, R. (1979). *Remote Sensing and Image Interpretation*, John Wiley & Sons, London.

Lo, C.P. (1986). *Applied Remote Sensing*, Longman, London.

Mather, P.M. (1989). *Computer Processing of Remotely Sensed Images: An Introduction*, John Wiley & Sons, Chichester.

Ritchie, W., Wood, M., Wright, R. and Tait, D. (1988). *Surveying and Mapping for Field Scientists*, Longman, London.

Rudd, R. (1975). *Remote Sensing – A Better View*, Duxbury Press, North Scituate, MA.

Sabins, F. (1987). *Remote Sensing – Principles and Interpretation*, Freeman, New York.

Useful Web Addresses

GIS in School Education

GIS: A Sourcebook for Schools Website
http://www.abdn.ac.uk/gis_school/index2.htm

AGI GIS in School Education Special Interest Group
http://www.gis-education.com

GIS in Schools Publication
http://www.agi.org.uk/pages/freepubs/gisinschoolsabs.html

Holgate School
http://homepages.tcp.co.uk/~heinrich/holg.htm

The National Curriculum
http://www.nc.uk.net/

Using GIS in Secondary Education: Curriculum, Implementation and Results
http://www.ncgia.ucsb.edu/conf/gishe96/program/ramirez.html

ESRI Schools and Libraries
http://www.esri.com/k-12

About GIS K-12 Education
http://gis.about.com/science/gis/cs/k12education/index.htm

National Geographic Kids
http://www.nationalgeographic.com/kids/index.html

The Globe Program
http://archive.globe.gov

USGS Rocky Mountain Mapping Centre – GIS in Education
http://rockyweb.cr.usgs.gov/public/outreach/giseduc.html

ESRI User Conference 1999 Paper
http://www.esri.com/library/userconf/proc99/proceed/papers/pap202/p202.htm

Intergraph Schools
http://www.intergraph.com/schools/

Introduction to GIS

GIS Primer - The Fundamental Concepts
http://www.innovativegis.com/education/primer/concepts.html

Introduction to GIS and Spatial Data
http://www.king.ac.uk/geog/gis_for_teachers/

Introduction to GIS
http://www.umass.edu/masscptc/gis.html

GIS Software and Data on the Web

ESRI MapExplorer
http://www.esriuk.com/products/Products.asp?pid=37

ESRI ArcExplorer
http://www.esri.com/software/arcexplorer/index.html

GeoCommunity
http://software.geocomm.com/

Palm Beach County
http://www.palmbeach.k12.fl.us/maps/gis/GISINFO.HTM
http://www.palmbeach.k12.fl.us/maps/textver.htm

The ESRI Geography Network
http://www.geographynetwork.com

The Data-Store
http://www.data-store.co.uk

GIS Materials and Tutorials

NCGIA ArcView-Based Learning Modules
http://www.ncgia.ucsb.edu/education/projects/SEP/avmodule.html

NCGIA Secondary Education Project Materials
http:// www.ncgia.ucsb.edu/education/projects/SEP/seppubs.html

Ordnance Survey Northern Ireland - Geographic Information System (GIS)
http://www.osni.gov.uk/multigis.htm

The Geographer's Craft
http://www.colorado.edu/geography/gcraft/contents.html

Taylor and Francis GIS and Remote Sensing Resource Centre
http://www.gis.tandf.co.uk/

Geography, GIS and Maps

Ordnance Survey
http://www.ordnancesurvey.gov.uk

National Geographic Maps
http://www.nationalgeographic.com/maps/index.html

MapZone
http://www.mapzone.co.uk/main.html

Remote Sensing

The Remote Sensing Core Curriculum
http://research.umbc.edu/~tbenja1/index.html

Canada Centre for Remote Sensing Education Page
http://www.ccrs.nrcan.gc.ca/ccrs/eduref/educate.html

National Learning Network for Remote Sensing
http://www.nln.met.ed.ac.uk

WWW Virtual Library: Remote Sensing
http://www.vtt.fi/aut/rs/virtual/

Taylor and Francis GIS and Remote Sensing Resource Centre
http://www.gis.tandf.co.uk/

GIS Community

National Council for Geographic Education
http://www.ncge.org/activities/events/gis.html

National Centre for Geographic Information & Analysis
http://ncgia.ucsb.edu/

GISLinx
http://www.gislinx.com

The GIS Portal
http://www.gisportal.com

GEOPlace
http://www.geoplace.com

The GIS Café
http://www.giscafe.com

The Advisory Unit – Computers in Education
http://www.advisory-unit.org.uk

GeoCommunity
http://www.geocomm.com

Index

GIS: A Sourcebook for Schools